Student Solutions Manual

for

Organic Chemistry
A Guided Inquiry

for Recitation, Volumes 1 & 2

a process oriented guided inquiry learning course

Andrei Straumanis

© 2012 Brooks/Cole, Cengage Learning

ALL RIGHTS RESERVED. No part of this work covered by the copyright herein may be reproduced, transmitted, stored, or used in any form or by any means graphic, electronic, or mechanical, including but not limited to photocopying, recording, scanning, digitizing, taping, Web distribution, information networks, or information storage and retrieval systems, except as permitted under Section 107 or 108 of the 1976 United States Copyright Act, without the prior written permission of the publisher.

For product information and technology assistance, contact us at
**Cengage Learning Customer & Sales Support,
1-800-354-9706**

For permission to use material from this text or product, submit all requests online at **www.cengage.com/permissions**
Further permissions questions can be emailed to
permissionrequest@cengage.com

ISBN-13: 978-1-111-57817-6
ISBN-10: 1-111-57817-6

Brooks/Cole
20 Davis Drive
Belmont, CA 94002-3098
USA

Cengage Learning is a leading provider of customized learning solutions with office locations around the globe, including Singapore, the United Kingdom, Australia, Mexico, Brazil, and Japan. Locate your local office at: **www.cengage.com/global**

Cengage Learning products are represented in Canada by Nelson Education, Ltd.

To learn more about Brooks/Cole, visit
www.cengage.com/brookscole

Purchase any of our products at your local college store or at our preferred online store
www.cengagebrain.com

Printed in the United States of America
1 2 3 4 5 6 7 14 13 12 11

How to Use This Solutions Manual

This volume contains the worked-out solutions to all the **Extend Your Understanding Questions** and **Confirm Your Understanding Questions** in both *Volume 1* and *Volume 2* of *Organic Chemistry: A Guided Inquiry for Recitation*.

The following are some tips for using this manual responsibly.

Make a thorough effort to answer a question *before* consulting this Solutions Manual.

When a student reads a question and doesn't know the answer, the tendency is to immediately look at the answer key. This hurts the learning process since most learning takes place as you try to figure out your own valid answer to a question by reviewing the ChemActivity, talking with others, reading the textbook, asking questions, etc.

If you have ever waited for tomorrow's paper to check your crossword puzzle answers, you know that the expert answers in this book will be far more meaningful to you if you have already spent time struggling to come up with your own *inexpert* answer. If you are pressed for time and go straight to the solutions manual, you will miss out on the most important part of the learning process, and fail to fully understand the answer given. It would be like filling in the crossword puzzle using the answers as reference.

If you can't answer the homework questions (without the key) you need to go back and review the activity.

It is much more important to know *if* you can answer the homework questions (so you know if you are ready for the quiz/exam), than for you to memorize the answers to these specific questions.

Many students find that consulting this manual too frequently ruins the diagnostic function of homework. This is because, in checking the answer to a question, you may accidentally see a structure or phrase that gives you a big hint about how to answer the next question.

Why are there no answers to the in-class Construct Your Understanding Questions?

You may have noticed that answers to the in-class Construct Your Understanding Questions are not provided. These questions are designed to be the focus of student discussion in (and after) class. If the instructor provided answers to these questions, your discussions would end prematurely, short-circuiting a key part of the learning process.

Regarding the in-class Construct Your Understanding Questions, many students at first wonder...

> *Wouldn't it be easier for students to learn the "right" answers if the instructor posted them?*

Of course it would be easier if the goal was simply to have students memorize answers to certain questions. However, it turns out that memorized answers are quickly forgotten and do not help you answer harder, more conceptual exam questions on the same topic.

On the other hand, **wrestling with a question and coming up with your own valid answer is a fantastic way to learn a topic**, and the best preparation for further study via homework and lectures in preparation for the exam.

Without an answer key, how do I figure out if my answer is valid?

Most of the learning in this class will take place while you are trying to figure out if your answers to the Construct Your Understanding Questions are correct. Constantly ask yourself and your teammates… **"Does this make sense?"** and **"Does this fit with what I already know?"**

As stated above…

> If you do not understand the homework, there is a good chance that you
> did not get all that you were supposed to get from the ChemActivity.

If you suspect you are missing something, try explaining your current understanding to a group mate, teaching assistant, or instructor. If your understanding is weak, there is a good chance that the act of trying to explain the ChemActivity will help you find your error.

Self-assessment (deciding if you are correct) is not always easy. In fact, one of the hardest parts of being a scientist is generating conclusions from the available data and deciding if they are valid.

The good news is that the self-assessment process is much easier when you are part of an effective group. That is why you are strongly encouraged to work in groups during and outside of class, and why almost all scientists work in teams.

If several students agree on an answer it is usually valid. If you are still not satisfied with this answer, ask another group or the instructor. Do not be surprised if your instructor tries to help you figure out your own answer rather than simply telling you the correct answer. Your instructor is not being difficult; he or she is trying to prepare you for the exam, and for the real world where there is no answer key.

After you graduate, you will face questions like "What is the effect of this gene mutation?", "How can I prove that my client is innocent?", or "What do these lab data mean?" You will be the first person to ever face this specific challenge. The skills you learn in this course will help you answer the question, make the correct diagnosis, publish the result, cure the patient, win the case, etc..

Table of Contents

Solutions for Questions in Volume 1 — page

1	Lewis Structures	1
2	Resonance	6
3	Constitutional Isomers	10
4	Cycloalkane Stereoisomers	12
5	Alkene Stereoisomers	16
6	Addition to Alkenes	19
7	Radical Halogenation	23
8	Chiral Centers	28
9	Absolute Configuration (R/S)	30
10	One-Step Substitution (S_N2)	34
11	Two-Step Substitution (S_N1)	40
12	Two-Step Elimination (E1)	45
13	One-Step Elimination (E2)	48
14	Sorting out E1, E2, S_N1, S_N2	53
15	Retrosynthesis	55
16	Carbon (^{13}C) NMR	58
17	Proton (^1H) NMR	62
NW1	Naming Alkanes & Cycloalkanes	67
NW2	Naming Functional Groups	68

Solutions for Questions in Volume 2 — page

1	Aromaticity	69
2	Introduction to EAS	74
3	EAS Resonance Effects	77
4	EAS Competing Effects	85
5	EAS Synthesis Workshop	87

Solutions for Questions in Volume 2

		page
6	Organometallic Reagents	90
7	Nucleophilic Addition-Elimination	93
8	Carboxylic Acids & Derivatives	105
9	Acid Halides & Anhydrides	110
10	Enol & Enolate Nucleophiles	118
11	Aldol Reactions	125
12	Aldol Condensations	130
13	Claisen & Michael Reactions	137
14	Amines	147
15	*Carbon (^{13}C) NMR*	see pages 58-61
16	*Proton (^1H) NMR*	see pages 62-66
NW3	Naming Benzene Derivatives	153
NW4	Naming Carbonyl Compounds	154

ChemActivity 1: Lewis Structures

Extend Your Understanding Questions (to do in or out of class)

13. Complete each of the following Lewis structures by adding any missing formal charges.

14. *see above*

15. False. The last structure in the middle row has formal charges but is overall neutral. This type of molecule (overall neutral with formal charges) is called a *zwitterion*.

16. The middle structure in the top row is technically not a legitimate Lewis structure because carbon lacks an octet.

17. *see above*

18. The three carbons in the left box each have exactly three bonds and a +1 formal charge. The three carbons in the right box each have exactly three bonds, one lone pair, and a -1 formal charge.

19. *See below.*

Recognizing Formal Charges for C, N, O, and X

	+1	0	−1
C	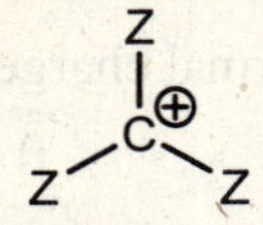 (carbocation structures shown) Note: The two other ways to draw a carbocation (shown on the previous page) are less common than this one.	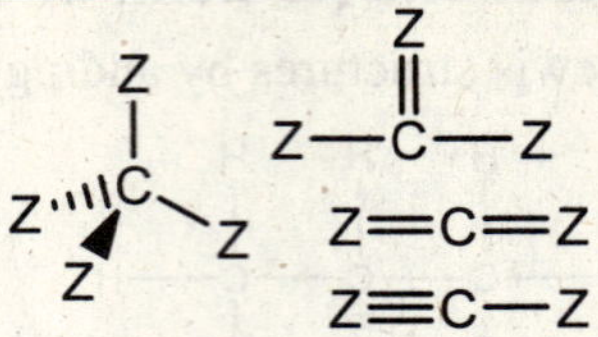 (three neutral C structures shown)	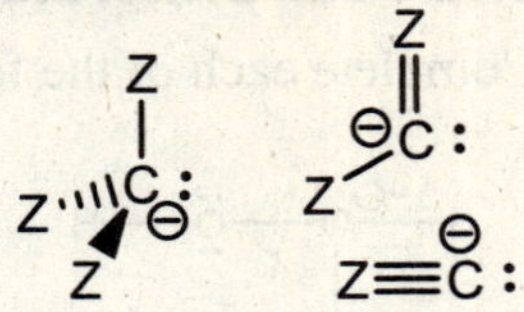 (two anionic C structures shown)
N	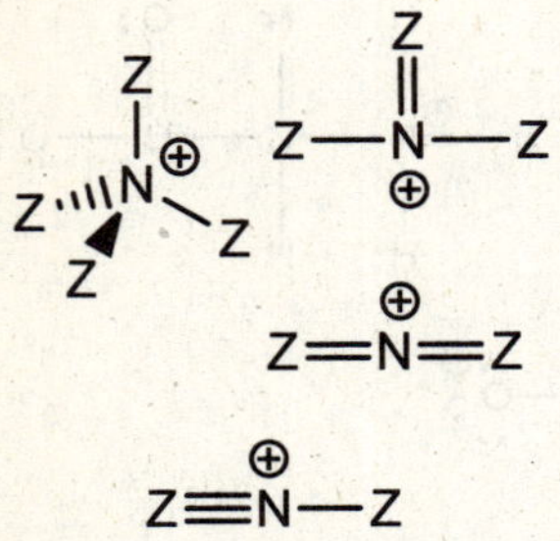 (structures shown) four ways (**draw the two that are missing**)	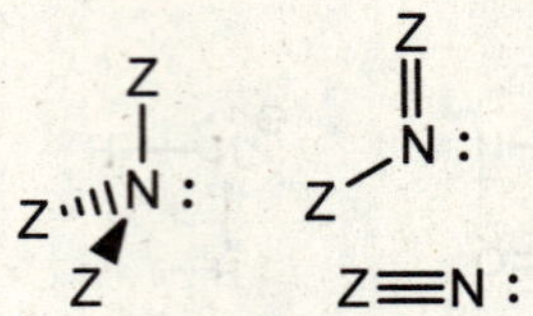 (structures shown) draw three ways	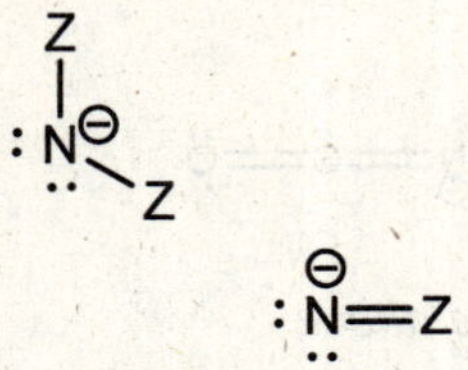 (structures shown) draw two ways
O	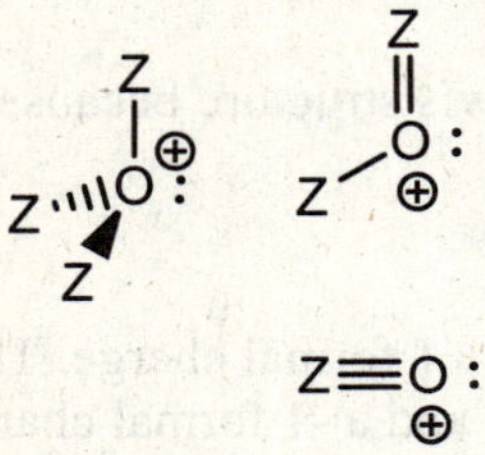 (structures shown) draw three ways	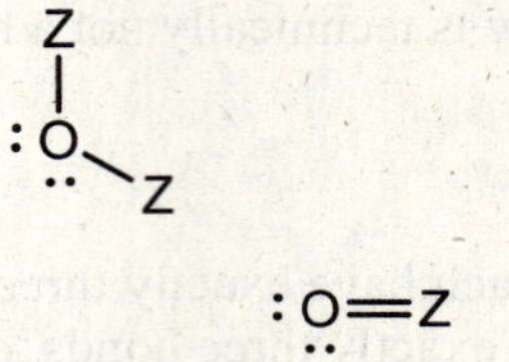 (structures shown) draw two ways	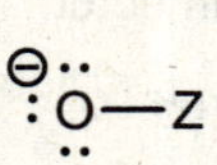 (structure shown) draw one way
X	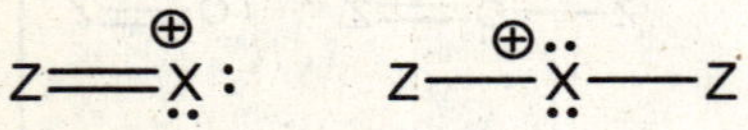 (structures shown) two ways (less common)	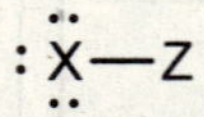 (structure shown) draw one way	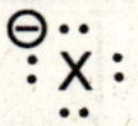 (structure shown) draw one way (an anionic atom)

20. *See above.*

21. Does the Lewis structure have the correct total number of electrons?
    Does each C, N, O and F have an octet (eight electrons in its valence shell)?
    Does each H have 2 electrons in its valence shell?
    Does this Lewis structure have the fewest possible non-zero formal charges?
    (Normally, the best Lewis structure will NOT have any formal charges > +1 or < −1.).

Confirm Your Understanding Questions (to do at home)

22. See below

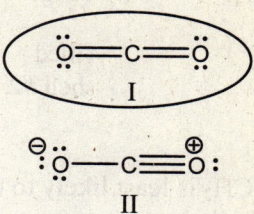

a. The circled Lewis structure of CO_2 is a better representation of how the electrons in the molecule are likely arranged based on the assumption that molecules adopt an electron arrangement with the fewest possible hot spots of + and – charge.

b. This structure also fits better with experimental data indicating that the two carbon-oxygen bonds of CO_2 are identical.

23. Shown below are two possible Lewis structures for the amino acid called glycine.

Structure I   Structure II

a. The ∠COH bond angle for Structure I is expected to be close to 109.5 degrees.

b. For Structure II, this angle is expected to be close to 120 degrees.

c. The estimate based on Structure I is more likely accurate since this is the better Lewis structure (fewer hot spots of + and – charge).

24. See below.

bent

25. See below.

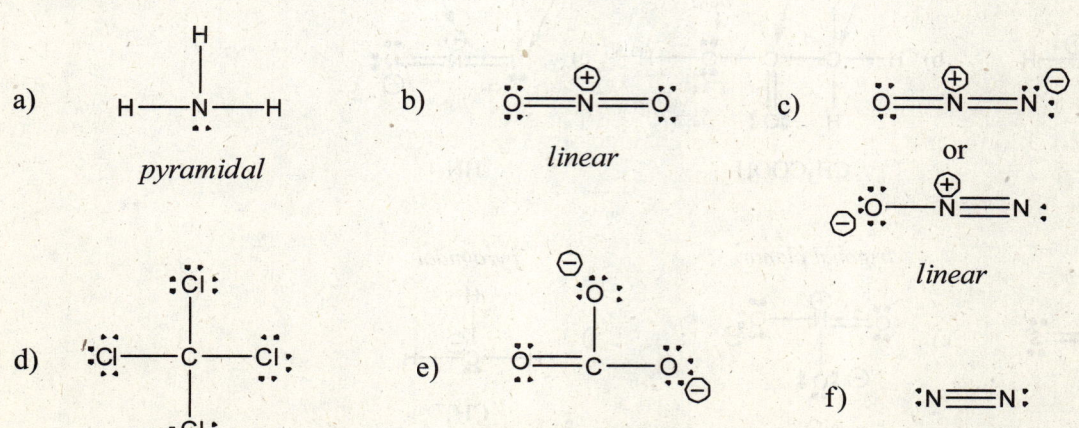

a) pyramidal

b) linear

c) or linear

d) tetrahedral

e) trigonal planar

f) Because of the way shape is defined, diatomics don't have an official shape.

26. Lewis structure of) the most commonly occurring ion...

a) :S:²⁻   b) :Mg: ²⁺ or Mg²⁺   c) :I:⁻   d) :O:²⁻
              filled   or empty
              shell #2   shell #3

27. CH₂ is least likely to exist since no legitimate Lewis structure can be drawn—Note that this species, called a carbene, can exist in certain special circumstances, but it is high energy and unstable. NO⁺ and HO⁻ have legitimate Lewis structures.

28. Lewis said nothing about boron or sulfur, only H, C, N, O and F. Third row elements (e.g. P or S) can expand their octet and boron often exists with six electrons in its valence shell. Carbon also can exist for very short periods of time with six valence electrons (this is called a carbene). Lewis' rules say nothing about carbocations and carbenes, which were discovered more recently.

29.

30. Missing formal charges are shown (along with shapes).

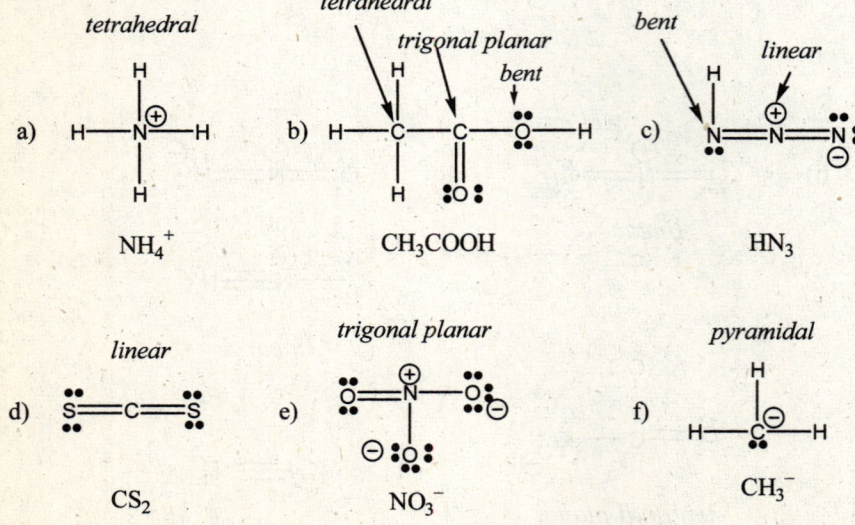

a) NH₄⁺ — tetrahedral
b) CH₃COOH — tetrahedral, trigonal planar, bent
c) HN₃ — bent, linear
d) CS₂ — linear
e) NO₃⁻ — trigonal planar
f) CH₃⁻ — pyramidal

31. Lewis structures (with shapes labeled).

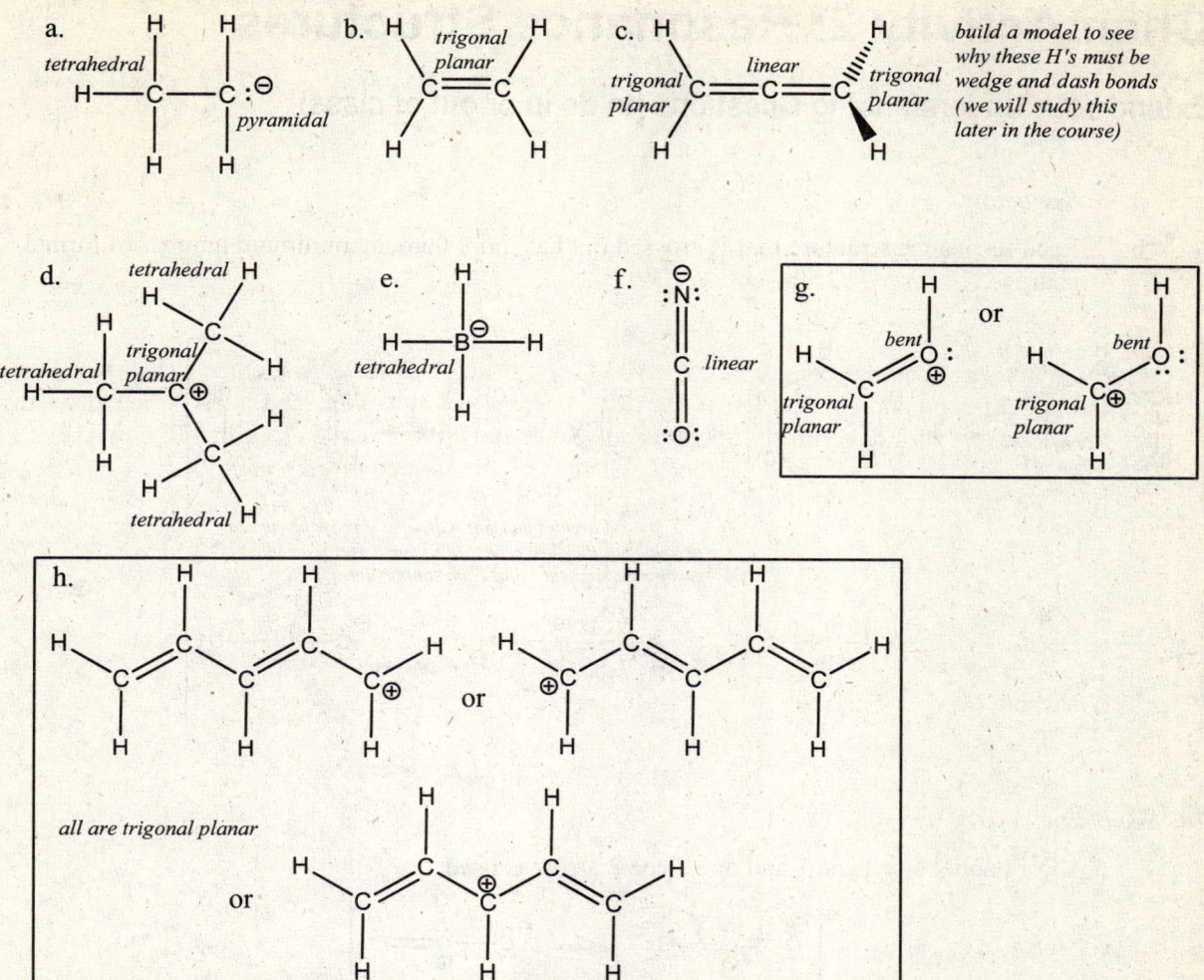

32. *See above.*

33. Lewis structure of carbon monoxide (CO).

34. The following is not a valid Lewis structure because carbon does not have an octet, AND it is not a carbocation, the one allowable exception to the octet rule.

ChemActivity 2: Resonance Structures

Extend Your Understanding Questions (to do in or out of class)

13.
   a. See below.

   b. Each resonance structure that is crossed out has more than the minimum number of formal charges.

The curved arrows marked with an X are valid curved arrows, but they are crossed out since they lead to a resonance structure that is not important.

14. See below.

   a. For ozone, one Type 1 and one Type 2 arrow is used.

   b. For benzene and toluene, only Type 3 arrows are needed.

   Benzene

   Toluene (also called Methylbenzene)
   a common benzene derivative

15. The five structures in the top row are considered **identical** representations of the same molecule because any one is enough to describe the molecule accurately; whereas the five structures in the bottom row are considered five **different resonance structures** because all five are needed to describe where the charge lies on this molecule. That is, if you use only one of the structures in the bottom row, this implies that the -1 charge is localized on a given carbon. You need all five to represent that the charge is distributed among the five carbons.

16. It is very common for students to assume (incorrectly) that a resonance structure that looks like the mirror image (or rotation) of another resonance structure is redundant, and therefore unnecessary. This is incorrect. See answer to Question 15, above.

Confirm Your Understanding Questions (to do at home)

17. Resonance structures of the molecule nitromethane (H_3C-NO_2).

18. Only the middle structure has more than one resonance structure.

19. *See below (continued on next page)*

Volume 1, ChemActivity 2: Resonance Structures

20. Resonance structures that are NOT important are crossed out.

21. *See below (Continued on next page.)*

The two resonance structures above are NOT the same, but to save time both RS's are usually shown using the one structure at right. To be complete this set would use the NO_2 shorthand or show 4 more resonance structures with a negative charge on the other oxygen of the NO_2 group.

22. It is NOT possible to draw a resonance structure of nitrate ion ($NO_3^{\ominus}$) that has only one formal charge.

23. a/c. *see below.* b. Use of the Type 3 arrow moved the negative charge past one of the carbons that (according to the full resonance description) is expected to hold a partial negative.

24. The complete set of resonance structures for the anion..

25. a. *see below.* b. You can think of Product A as arising from the Resonance Structure #1, and Product B arising from RS#2. It follows then that even though RS#1 and RS#3 are different, they will both give Product A. This logic leads you to conclude that the ratio of A:B will be 2:1. It turns out that this system is more complicated, but we have not yet learned the other factors that affect this system.

Resonance Structure 1   Resonance Structure 2   Resonance Structure 3   Product B

ChemActivity 3: Constitutional Isomers

Extend Your Understanding Questions (to do in or out of class)

13. The degree of unsaturation is written below each structure.

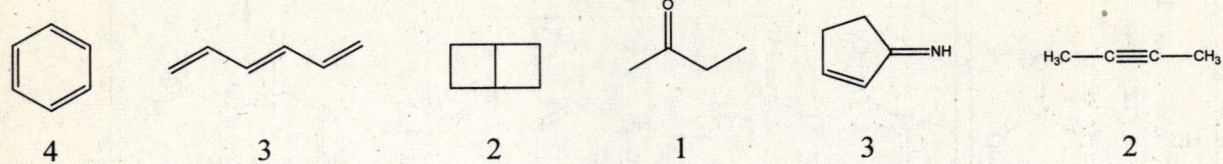

14. The degree of unsaturation for a molecule with molecular formula C_7H_8NOBr is 4.

15. The last molecule cannot have the molecular formula C_7H_8NOBr because it has 5 degrees of unsaturation.

16. A molecule with molecular formula $C_6H_4O_2$ has 5 degrees of unsaturation. A possible structure is…

Confirm Your Understanding Questions (to do at home)

17. A branched alkane with n carbons will always have the same molecular formula (C_nH_{2n+2}) as a straight-chain alkane with n carbons because making a new branch on the chain changes a CH_2 to a CH, and this exactly balances the fact that making a branch creates a new terminal CH_3 on the molecule from a CH_2.

18.

The constitutional isomer that is missing from column 1 of Model 2 is…

19. The constitutional isomers missing from column 2 in Model 2 are…

The possibilities shown ignore stereoisomerism (which is introduced in a later section).

20. No constitutional isomers are missing from Column 3 in Model 2.

21. Constitutional isomers with the formula $C_5H_{11}F$.

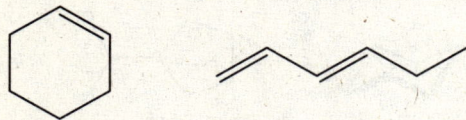

22. Six-carbon alkene (containing only C and H) with one ring and one double bond is…

    a. A constitutional isomer of this structure <u>with no rings</u> is shown above.

    b. The statement that "In terms of molecular formula, a ring is equivalent to a double bond" refers to the fact that introducing a ring into a structure has the same effect on the molecular formula as introducing a pi bond—to reduce the number of H's by two.

23. The purpose of Question 8 is to remind you that rotating a single bond does not change the identity of a molecule. That is, hexane can be represented many different ways, but it is still hexane. Another way to think about it is that, at a given moment, a bottle of hexane (which is a low-boiling liquid similar to gasoline) contains molecules of hexane in every possible conformation. By convention, chemists usually represent a molecule in its simplest or lowest potential energy conformation. For hexane this is the representation at the top left corner of Model 2. For cyclohexane this a flat looking hexagon. (In a future ChemActivity we will learn that, in fact, cyclohexane is not flat, but for simplicity it is often represented that way.)

ChemActivity 4: Cycloalkane Stereoisomers

Extend Your Understanding Questions (to do in or out of class)

14. 1 & 6 (identical); 2 & 3 (configurational stereoisomers); 2 & 4 (constitutional isomers); 4 & 5 (different molecular formula); 5 & 8 (constitutional isomers); 2 & 7 (constitutional isomers); 8 & 9 (conformers); (the following pairs are all examples of constitutional isomers…) 7 & 10; 11 & 12; 11 & 13; 4 & 14; 11 & 2; 11 & 3; 12 & 2; 7 & 14 ; as well as, 12 & 13; 12 & 3; etc.

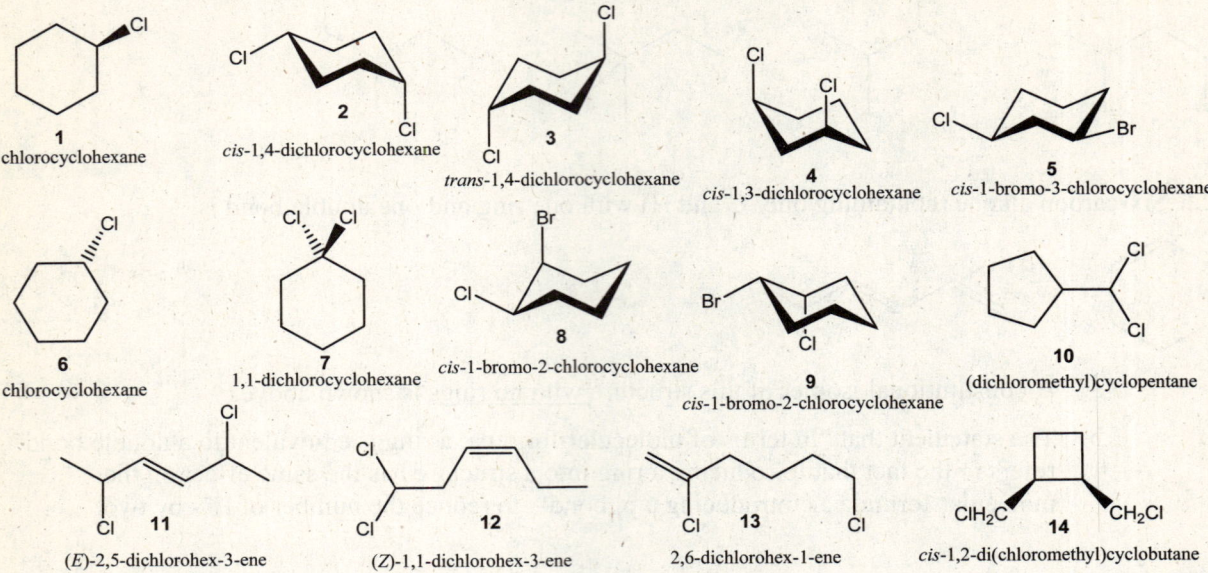

15. *See below.*

Generalized name	*cis* configurational stereoisomer	*trans* configurational stereoisomer
1,2-dimethyl-cyclohexane		lower PE
1,3-dimethyl-cyclohexane	lower PE	
1,4-dimethyl-cyclohexane		lower PE

16. *See above.*

Confirm Your Understanding Questions (to do at home)

17. This student appears to believe that "*cis*" means "same side" in terms of being on the same side of the ring (e.g. C_1 and C_2) and that "*trans*" means "across" in terms of being across the ring from one another (e.g. C_1 and C_3). This student is not following the algorithm for determining *cis* and *trans* for a ring in which the plane of the ring is first identified, and groups are then categorized as being on the same side of this plane (*cis*) or opposite sides of this plane (*trans*).

18. The following information is conveyed by each part of this name:

    <u>cis</u>   the 3D relationship of the two methyl groups

    <u>1,2-</u>   the location of the two methyl groups on the main chain

    <u>di</u>   there are two methyl groups (this seems redundant since 1,2- tells you there are two methyl groups, but it's still required).

    <u>methyl</u>   the identity of the groups on the ring

    <u>cyclo</u>   that the main chain is a ring

    <u>pentane</u>   that the main chain has five carbons (and the ending tells you there are no pi bonds)

19. See previous page.

20.

Br——⬡······Cl	⬡^{Cl}⁄_{Cl}	HO—⬡······CH₂CH₃	⬡—NH₂ with H₂N
trans	neither	*trans*	*cis*

Br——⬡——Cl		HO—⬡——CH₂CH₃	⬡······NH₂ with H₂N
cis		*cis*	*trans*

21. It is true that if you perform a chair flip on *cis*-1,4-dichlorocyclohexane, the result is *still* called *cis*-1,4-dichlorocyclohexane.

22. During a cyclohexane ring chair flip…

    a. …all axial positions become equatorial, all equatorial positions become axial, BUT all up groups remain up and all down groups remain down.

    b. …a group that was <u>up</u> and in an <u>axial</u> position will be up and in an equatorial position after a chair flip has occurred.

23. Fill in the blanks: *cis*-1,3-dibromocyclohexane has two different chair conformations: one with both Br groups in **equatorial** positions and one with both Br groups in **axial** positions.

24. Structure 5 in Question 14 is lower in potential energy because both groups are in equatorial positions, which places them farther away from the ring carbons and reduces steric repulsion.

25. *Trans*-1-*tert*-butyl-4-methylcyclohexane is lower in potential energy than *cis*-1-*tert*-butyl-4-methylcyclohexane because the former has a conformation available in which both groups are in equatorial positions at the same time (so it will almost always adopt this conformation).

26.

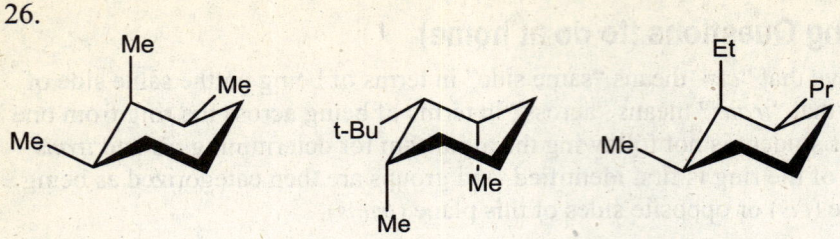

27.

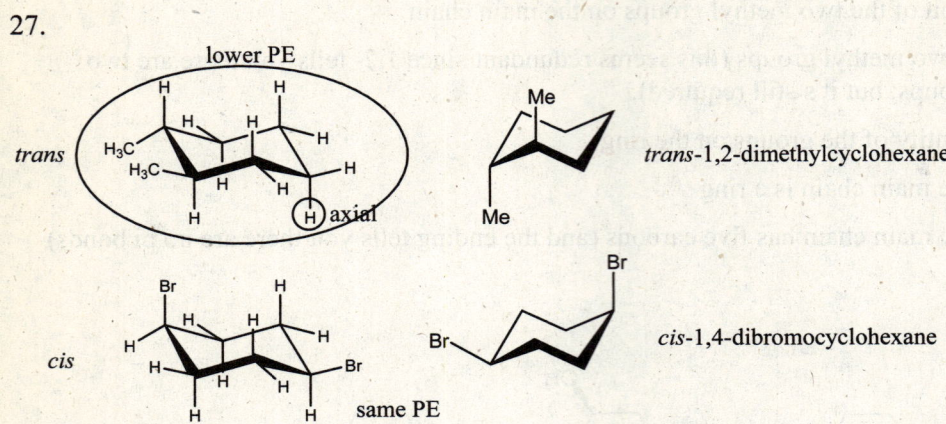

28.

this methyl group should be pointing straight down if it is axial or as shown below if it is equatorial

can't tell if the OH group is up or down, axial or equatorial

R_2 should be pointing slightly down; R_3 should be pointing slightly up

This Cl should be pointing straight down if axial or out if equatorial

can't tell if the Pr group is up or down, axial or equatorial

This OH should be pointing straight up if axial or out if equatorial

Corrected versions...

or

or

or

or other possibilities

29.
- a. X and H1 (neither)
- b. X and Z (trans)
- c. Y and Z (cis)
- d. H2 and H3 (cis)
- e. Z and H3 (cis)
- f. Z and H4 (trans)
- g. Y and H3 (cis)
- h. H3 and H4 (neither)

30. The t-butyl group is so large that it is nearly always found in an equatorial

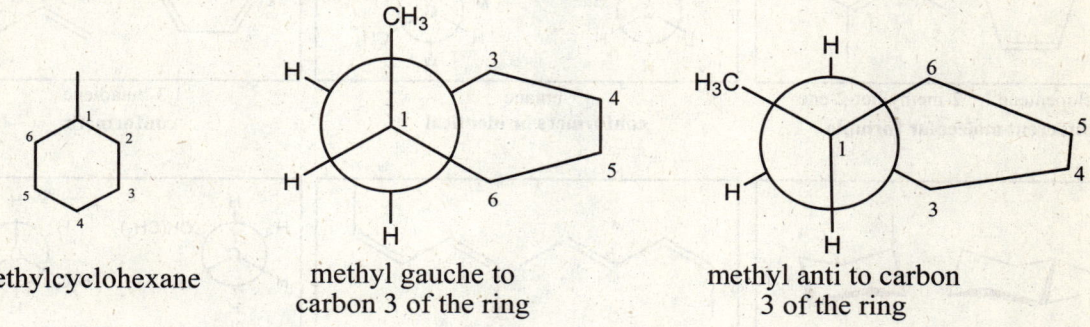

31. Newman projections of methylcyclohexane in the chair conformation:

methylcyclohexane

methyl gauche to carbon 3 of the ring

methyl anti to carbon 3 of the ring

- a. When the methyl group is in an **equatorial** position, the molecule is in its lowest potential energy conformation.
- b. *see above*
- c. In general, anti is a lower PE conformation.
- d. The ring carbons (C_3 and C_6) will always be gauche to one another because the ring structure prevents the C_1C_2 bond from being rotated beyond the rotation involved in a chair flip. Placing the next largest group on C_1 (in this case methyl) anti to C_3 gives it the most space (lowers the steric hindrance). This question is designed to illustrate that putting the methyl group anti to C_3 places it in an equatorial position.

ChemActivity 5: Alkene Stereoisomers

Extend Your Understanding Questions (to do in or out of class)

7. Indicate the relationship between each pair. Choose from: **configurational stereoisomers, conformers or identical, constitutional isomers,** or **different formulas** (use *each* at least twice).

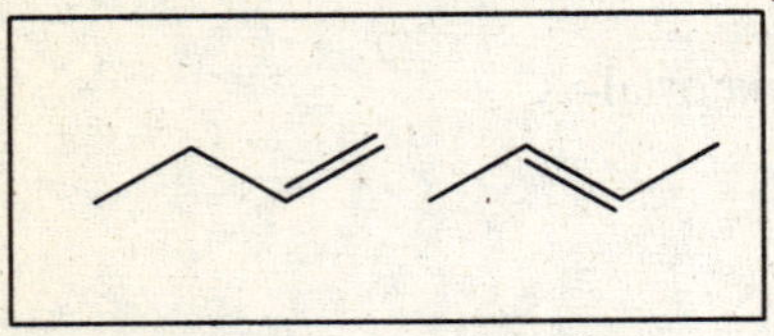

but-1-ene E-but-2-ene **constitutional isomers**	butane **conformers**	trans-1,3-dimethyl-cyclopentane cis-1,3-dimethyl-cyclopentane **configurational stereoisomers**	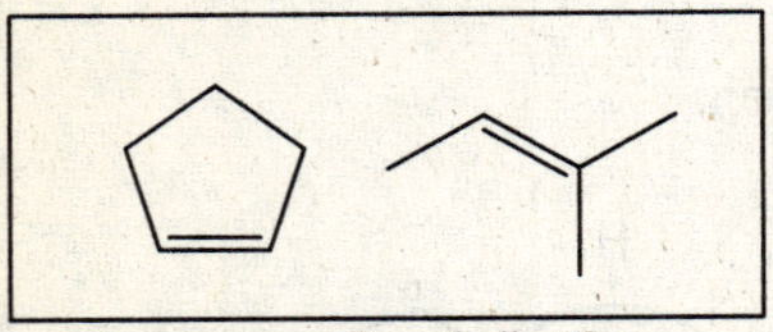
cyclopentene 2-methylbut-2-ene **different molecular formula**	butane **conformers or identical**	1,3-butadiene **conformers**	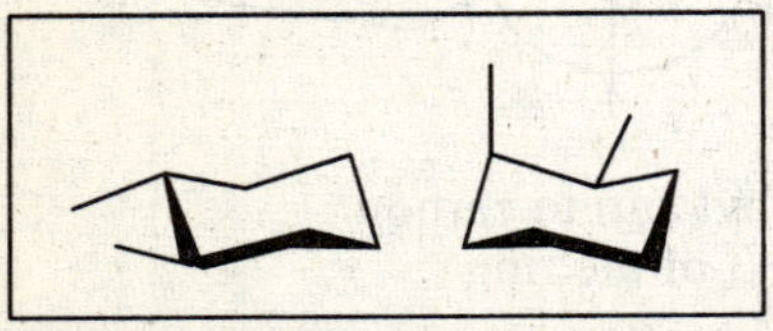
trans-1,2-dimethyl-cyclohexane cis-1,2-dimethyl-cyclohexane **configurational stereoisomers**	(E,Z)-2,4-hexadiene (Z,Z)-2,4-hexadiene **configurational stereoisomers**	2-methylbutane **conformers or identical**	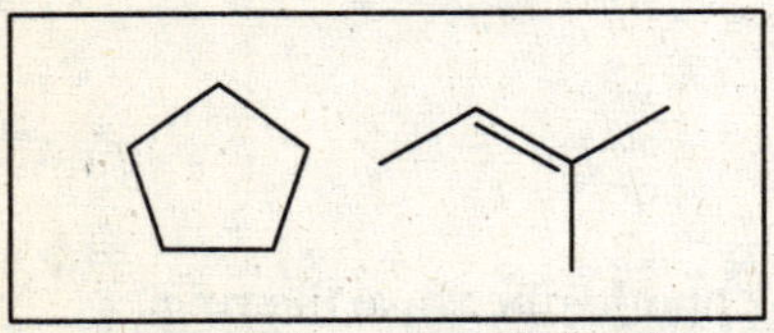
cyclopentane 2-methylbut-2-ene **constitutional isomers**	E-3-methylpent-2-ene Z-3-methylpent-2-ene **configurational stereoisomers**	2-methylbut-2-ene **conformers or identical**	

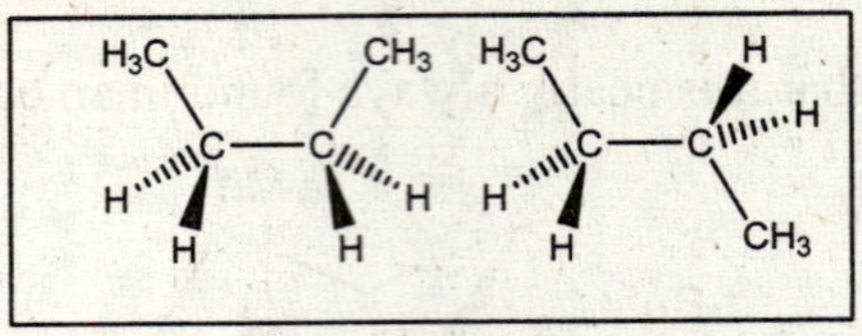

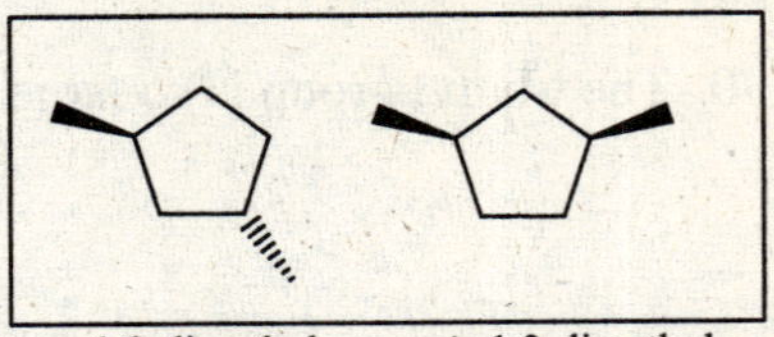

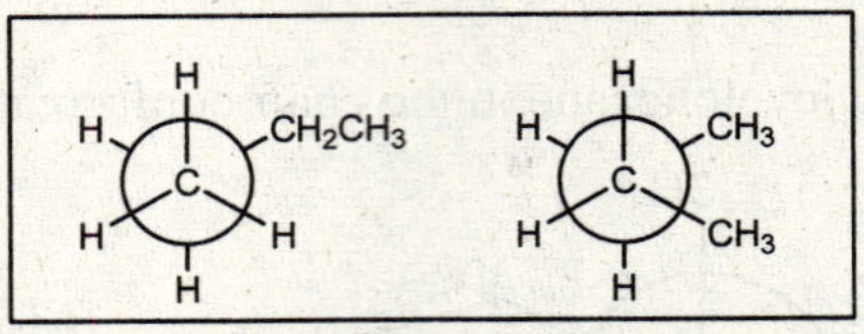

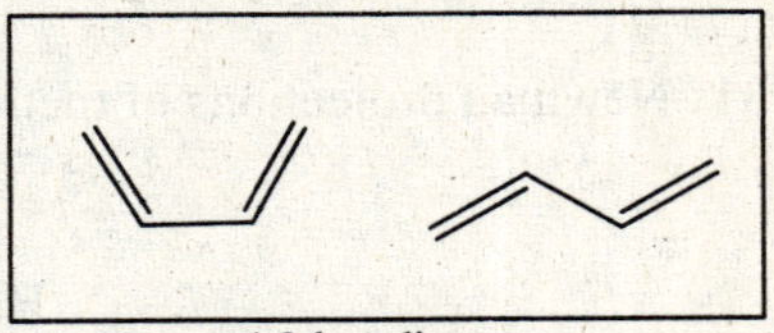

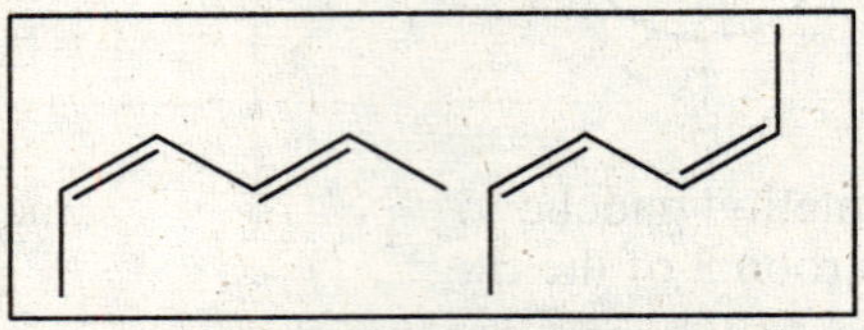

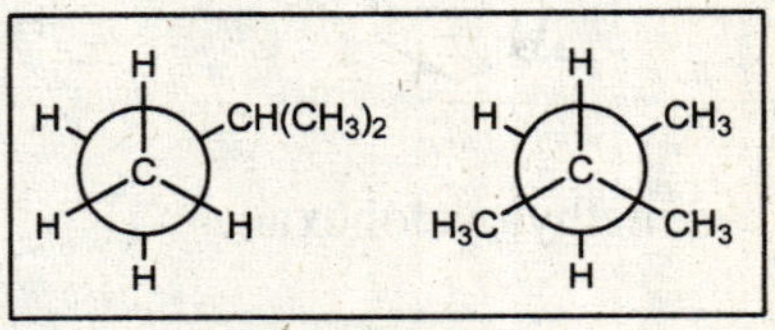

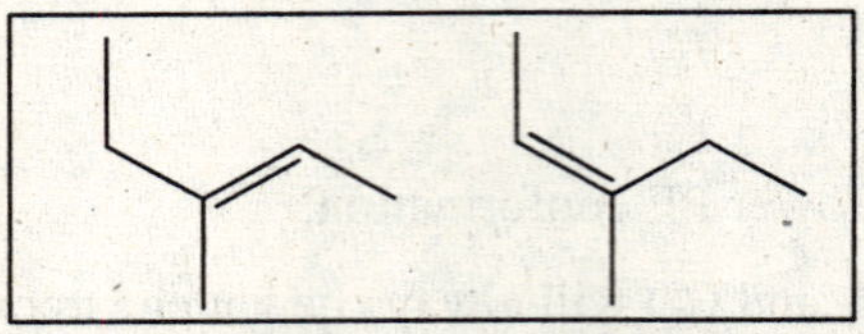

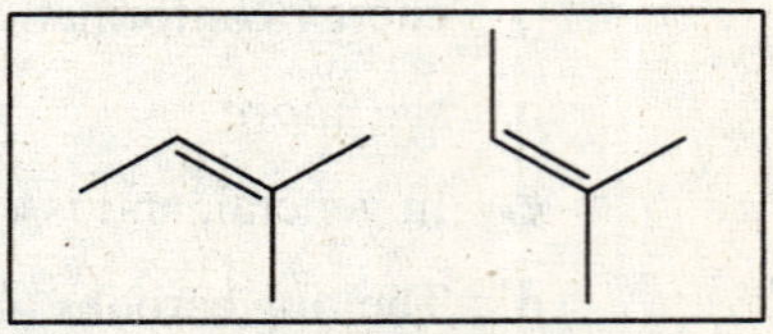

8. *See above.*

Confirm Your Understanding Questions (to do at home)

9. Draw a skeletal representation of *Z*-2-hexene and *E*-2-hexene.

10. I doest not make sense to specify either Z- or E- for but-1-ene while you must specify Z- or E- for but-2-ene because there are two identical groups (H's) attached to one end of 1-butene so it is neither E nor Z.

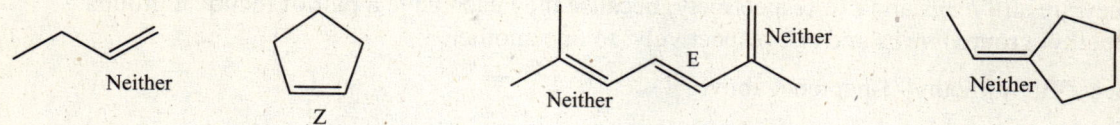

11. Each double bond is labeled E (trans), Z (cis), or neither.

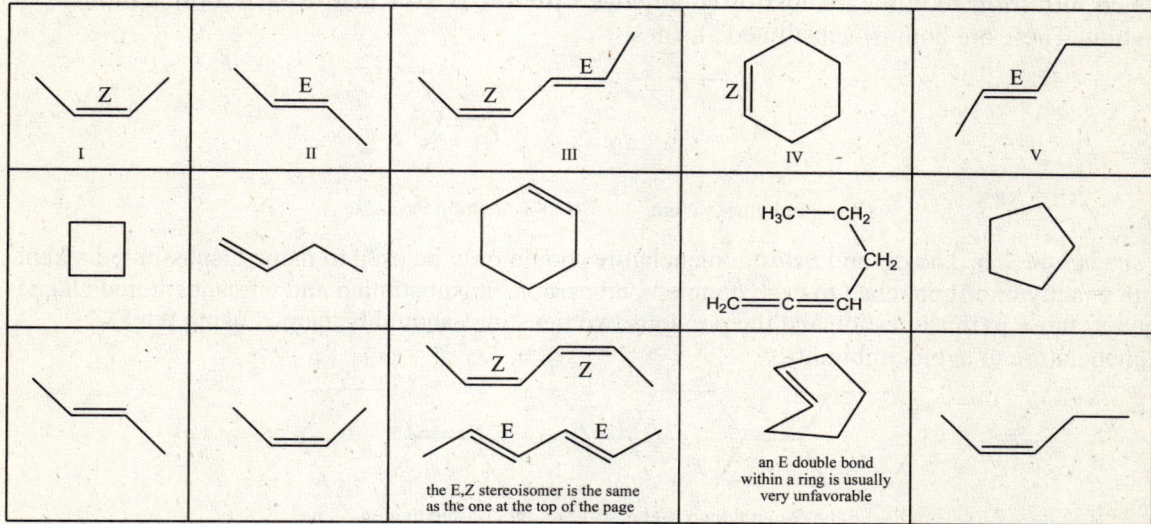

12. See below (continued on the next page).

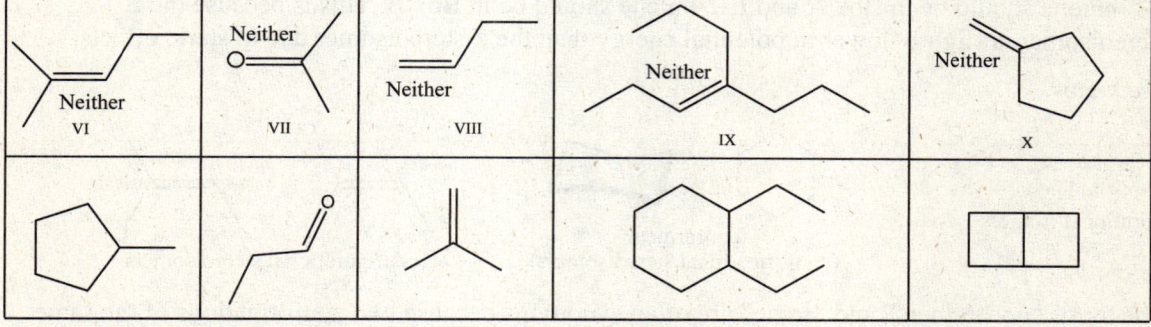

a. *See above.* Note that no structure in the second row has a configurational stereoisomer.

b. A terminal double bond is one that has a CH_2 group. Since the two H's constitute two identical groups attached to the same carbon in the double bond is neither E or Z.

c. Constitutional isomers

13. Skeletal structure of *E*-3,4-dimethyl-3-heptene and *Z*-3,4-dimethyl-3-heptene...

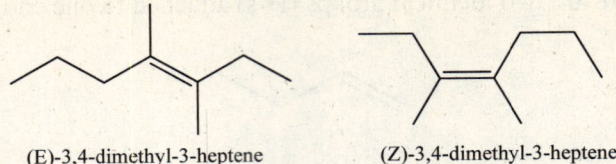

(*E*)-3,4-dimethyl-3-heptene   (*Z*)-3,4-dimethyl-3-heptene

   a. Though the largest groups on either end of the double bond in these molecules are not identical, they are still *trans* and *cis*, respectively, because they each have a pair of identical groups (methyl groups) *trans* and *cis*, respectively, to one another.

   b. See *Z*-3,4-dimethyl-3-heptene, above.

14. The name *cis*-3-methylpent-2-ene does not make it clear whether you intend the methyl groups to be *cis* to each other or the two largest groups to be *cis* to each other. **In general, you should only use *cis* and *trans* to name disubstituted alkenes with one H attached to each double bond carbon.** These are both tri-substituted alkenes.

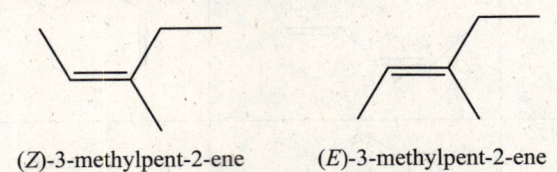

(*Z*)-3-methylpent-2-ene   (*E*)-3-methylpent-2-ene

15. a. *see below*   b. The *cis* and *trans* nomenclature should only be used to name disubstituted alkenes with exactly one H attached to each double bond carbon. Trisubstituted and tetrasubstituted alkenes such as those in this question and the previous two questions should be named using the E/Z nomenclature to avoid ambiguity.

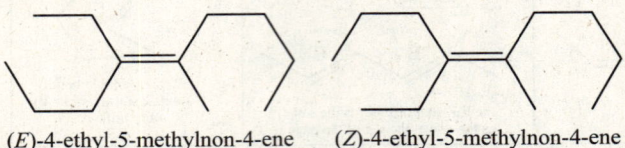

(*E*)-4-ethyl-5-methylnon-4-ene   (*Z*)-4-ethyl-5-methylnon-4-ene

16. *Z*-2-butene should be in Box A and *E*-2-butene should be in Box B. This is because the *E* stereoisomer is slightly lower in potential energy than the *Z* stereoisomer due to steric effects.

17. *See below.*

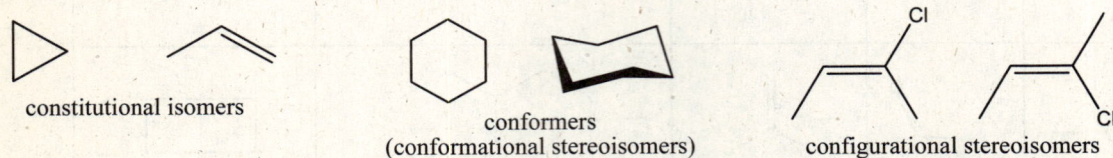

constitutional isomers   conformers (conformational stereoisomers)   configurational stereoisomers

18. The terms "conformers" and "same" are often synonyms because two conformations of the same molecule can be shown on paper, but they usually represent the same molecule with the same name, same physical properties, etc., and the two conformations could not be isolated under normal conditions in the laboratory. (Note that at very low temperature or with special molecules with very large barriers to single bond rotation, two different conformers may be isolated and studied.

19. Five constitutional isomers with formula C_6H_{12} that utilize a 2-methylpentene backbone are...

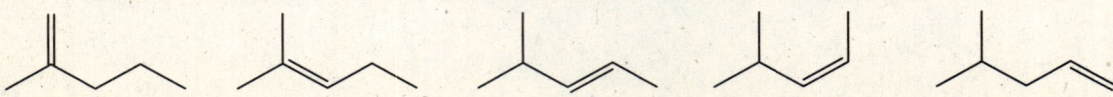

ChemActivity 6: Addition to Alkenes

Extend Your Understanding Questions (to do in or out of class)

18. (Both boxes) Tertiary carbocations are lower in potential energy than secondary carbocations.

19. (Box 1) The C-H bond (at the tail of the arrow) breaks, and the electrons in the bond form a new bond from the H (indicated) to the carbon (at the head of the arrow).

    (Box 2) The C-C bond (at the tail of the arrow) breaks, and the electrons in the bond form a new bond from R_1 to the carbon (at the head of the arrow).

20. *See below.*

21. In Anti-Markovnikov addition of HBr to an alkene (Synth. Transf. 6.2) the Br ends up on the less substituted of the two alkene carbons (or the carbon that would make a less favorable location for a carbocation [or radical]). In Markovnikov addition of HX to an alkene (Synth. Transf. 6.1), the X (e.g. Br) ends up on the more substituted of the two alkene carbons (or the carbon that is a more favorable location for a carbocation).

Confirm Your Understanding Questions (to do at home)

22.

23. These products would form via less favorable (higher potential energy) carbocations. The tertiary carbocations in the mechanisms above are much more likely to form.

24. Reaction of HX (X = Cl, Br or I) with ethene would lead to a primary carbocation, and primary carbocations do not normally form.

25.

28.

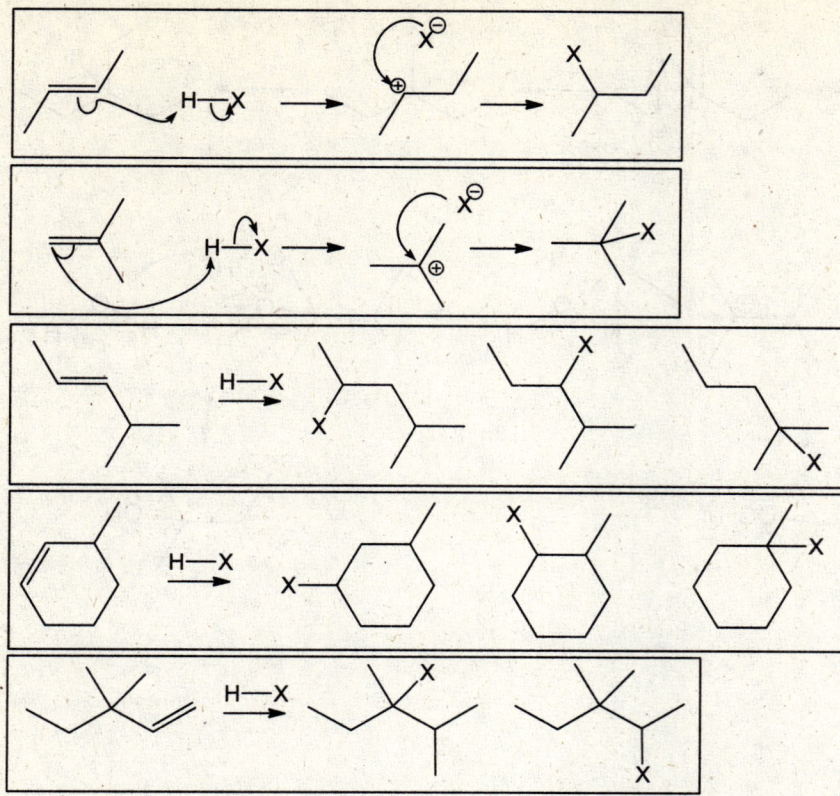

29. a. and b. (*see below*) c. The first rearrangement is downhill because the strain associated with a four member ring is much higher than the strain associated with a five-member ring. The 2nd rearrangement is a typical 2° to 3° carbocation rearrangement driven by the lower PE of a 3° carbocation.

30.

31.

32.

a.

b.

ChemActivity 7: Radical Halogenation

Extend Your Understanding Questions (to do in or out of class)

18.

Br–Br →(hν) Br•   •Br + H₃C–H → H–Br + •CH₃   •CH₃ + Br–Br → CH₃–Br + Br•

    initiation step   propagation step   propagation step

19. There is only a very small chance that two of the five "violently reactive fans" will get in a fight with one another. This is because each of these five jerks will pick a fight with a neighbor, and given the low number of jerks relative to the capacity of the stadium, this neighbor is very likely to be an ordinary citizen of the City of Brotherly Love, who is trying to enjoy the game in peace. This analogy is useful for understanding why a radical is most likely to make a bond to a non-radical (assuming the non-radical species are in much higher concentration in solution). In a reaction mixture there are billions of times more participants than in the stadium. For this reason, radicals eventually bump into each other in a reaction mixture, leading to a termination step.

20. Many students give the following answer to Question 18. The answer at the top of this page is much better than the one below since the step below showing collision of two radicals (though it gives the correct product) is much less likely to occur since it involves a collision between two low-concentration species.

Br–Br →(hν) Br•   •Br + H₃C–H → •CH₃ + Br• → CH₃–Br   termination step

21. a. The most prevalent species in the reaction mixture after step 2 is Br₂. b. See the last step in the mechanism at the top of the page.

22. See answers to Questions 18 and 20, above. Additional termination steps are shown below. (Note that a termination step does not have to yield the desired product to qualify as a termination step.)

Br• + •Br → Br–Br   termination step

•CH₃ + •CH₃ → H₃C–CH₃   termination step

Confirm Your Understanding Questions (to do at home)

23. If the Br radical reacts with the <u>most abundant species in the reaction mixture</u> (Br-Br), the products are identical to the reactants, and this is not very interesting.

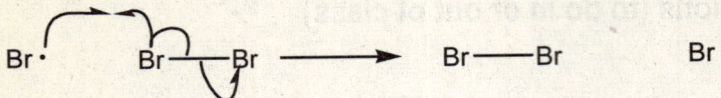

24. Each of the solvents on the left are suitable for radical halogenation reactions because they have no H's that can be replaced in such a reaction (the H's on benzene are very unreactive for reasons we will discover later). The solvents on the right will, themselves, undergo radical halogenation, using up halogen reagent.

25.

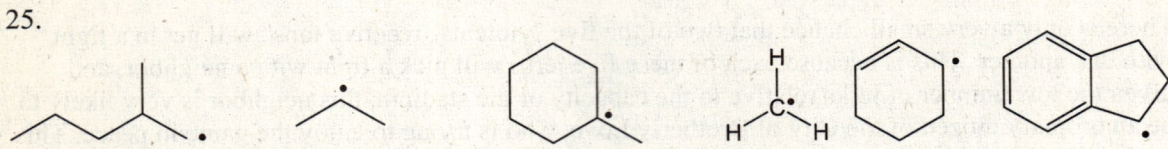

26. a. *See below*. b. The benzylic radical is much lower in potential energy than the radical shown in the question with a benzene ring and a radical on an atom two away from the ring because the former is highly resonance stabilized, but the latter is not. Note that, unlike carbocations, radicals (for reasons we will not discuss) do not rearrange. [If they could, you would expect the higher energy radical in this question to rearrange and become the lower energy benzylic radical.]

27. a. The radical on the left is lower in PE because it is resonance stabilized (see below).
    b. The radical on the left is an allylic radical.
    c. One resonance structure of an allylic carbocation is shown in the box below, right. Allylic carbocations, like allylic radicals, experience resonance stabilization. Radicals are also similar to carbocations in that more substituted radicals are lower in potential energy. One difference between carbocations and radicals is that carbocations can rearrange, but radicals cannot.

28. The middle of the three diagrams best tracks the products formed from a single initiation event (breaking apart of one Br_2 molecule) in **a** radical chain reaction.

29.

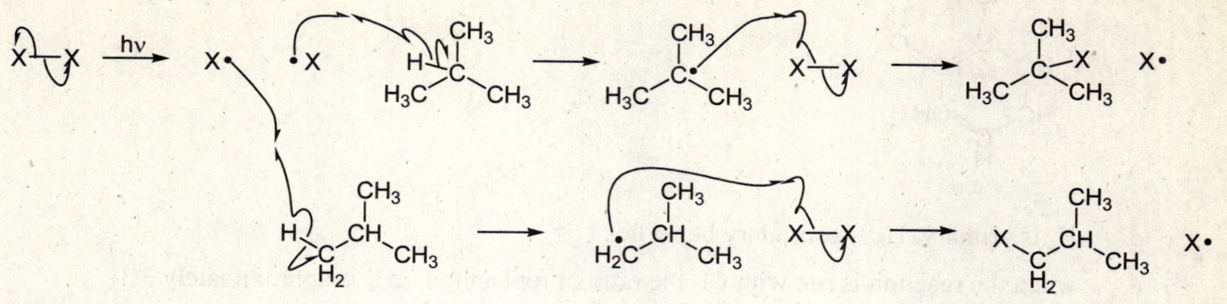

[Reaction mechanism scheme showing radical halogenation of isobutane: X-X undergoes hv to form 2 X•; X• abstracts H from (CH₃)₃CH to form tertiary radical (CH₃)₃C•, which reacts with X-X to give (CH₃)₃C-X and X•. Lower pathway: X• abstracts a primary H from (CH₃)₂CHCH₃ to form primary radical •CH₂CH(CH₃)₂, which reacts with X-X to give XCH₂CH(CH₃)₂ and X•.]

30. Referring to the two products shown above...

   a. An H is more likely to be replaced with a halogen in the order benzylic, allylic, 3°, 2°, 1°, methyl; however, only for bromine is selectivity for the more preferable H a large factor.

   b. The number of H's of a given type determines the number of chances a halogen radical has of replacing that H. This is a straight multiplicative factor. That is, there are nine times as many primary H's on the starting material above compared to tertiary H's. This means you must multiple the probability of replacing a given H by the number of H's of that type. For example. If X = Cl, the probability of replacing a primary H versus a tertiary H is 1:5. However, since there are nine time more primary H's, the ratio of the primary alkyl halide product to the tertiary alkyl halide product will be 9:5, with the primary product dominating.

   c. The identity of X (F, Cl, Br, or I) determines the speed and selectivity of the reaction. Fluorine is so fast as to be uncontrollable. Iodine is so slow that it is useless. Cl and Br are useful, but only Br is highly selective. (The selectivity of the reaction is the likelihood that a given type of H will be replaced.)

31. a. F_2 b. I_2 c. Br_2 d. Cl_2

32. a. The slight selectivity of Cl for a secondary benzylic position nearly counterbalances the fact that there are three primary benzylic H's (and only two secondary benzylic H's). Both of these would dominate over the third product.
   b. Since Cl is much less selective than Br, switching to Br will give a higher yield of the desired (first) product. The second product might be found in small amounts, but the third product is very unlikely to form given the strong preference of Br for benzylic positions.

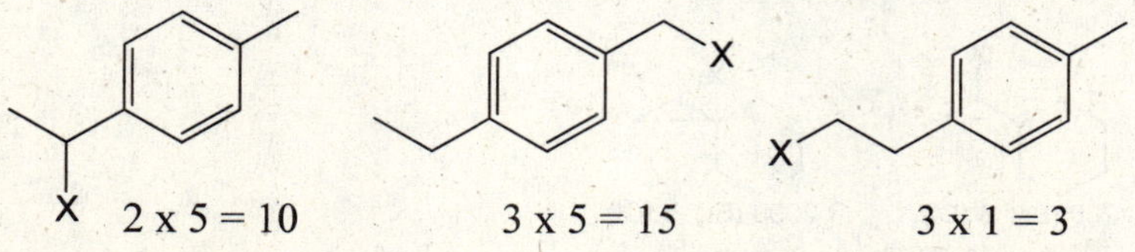

[Three structures shown: 1-(4-methylphenyl)-1-haloethane with X on benzylic carbon: 2 × 5 = 10; 4-ethylbenzyl halide (X on CH₂ attached to ring): 3 × 5 = 15; 2-(4-methylphenyl)-1-haloethane (X on terminal CH₂): 3 × 1 = 3]

33.
a., b. [structure with labeled hydrogens a, b, c, d on a cyclohexadiene with ethyl and methyl substituents]

c. H_a is primary. H_b is secondary benzylic.

d. when the reaction is run with Cl_2 the ratio of replacing H_a:H_b is approximately 3:10

e. when the reaction is run with Br_2 the ratio of replacing H_a:H_b is very small:very large

f. The reaction with Br_2 is much more selective because bromine selectively generates one specific product, whereas chlorine generates a similar amount of two different products.

34.

Br_2, hv: 6 (1.5%), 400 (98.5%)
Cl_2, hv: 6 (27%), 16 (73%)

Br_2, hv: allyl bromide
Cl_2, hv: allyl chloride

Tetralin + Br_2/hv: (99.9%) and (0.1%)
Tetralin + Cl_2/hv: similar

Ethylcyclohexane + X_2/hv:
- 200 (Br); 8 (Cl), 6%, 14.2%
- 400 (Br); 16 (Cl), 12%, 28.5%
- 400 (Br); 16 (Cl), 12%, 28.5%
- 3 (Br); 3 (Cl), <1%, 5%
- 200 (Br); 8 (Cl), 6%, 14.2%
- 2000 (Br); 5 (Cl), 62%, 9%

Neopentane + X_2/hv: only product for both

35.

36.

ChemActivity 8: Chiral Centers

Extend Your Understanding Questions (to do in or out of class)

16. Each molecule in the "Not Chiral" box has an **internal mirror plane** that cuts the molecule in half from left to right (as drawn).

17. Each carbon indicated with a "1" is considered to have **two identical groups** because going around the ring one way is identical to going around the other way. One way to think about this is to imagine that you are taking a walk around the ring. In either direction, on such a walk you would encounter exactly the same groups at the same point in your walk. In contrast, each carbon with an * has no identical groups.

18. Each * in Model 5 marks a chiral center with four different groups attached.

19. *See below*.

meso-tartaric acid   (+)-tartaric acid

achiral   achiral
meso   meso

20. *See above*. Each of the molecules above marked *meso* have chiral centers, but are not chrial.

21. There are no chiral centers in any of the molecules in Model 6.

22. Nither pair of molecules in Model 6 an identical pair based on the definition of identical. That is, neither can be superimposed on the other.

23. None of the molecules in Model 6 have an internal mirror plane.

24. All four molecules in Model 6 are chiral since they are not identical to their mirror images.

Confirm Your Understanding Questions (to do at home)

25.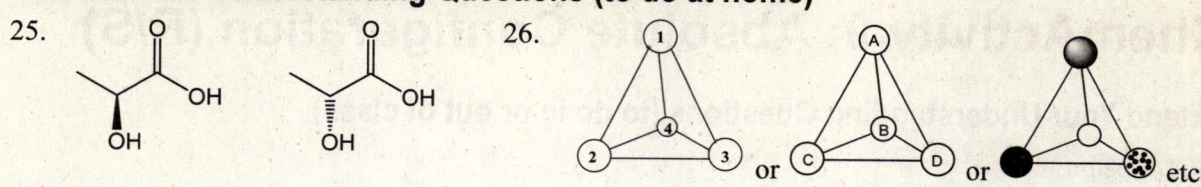

26.

27.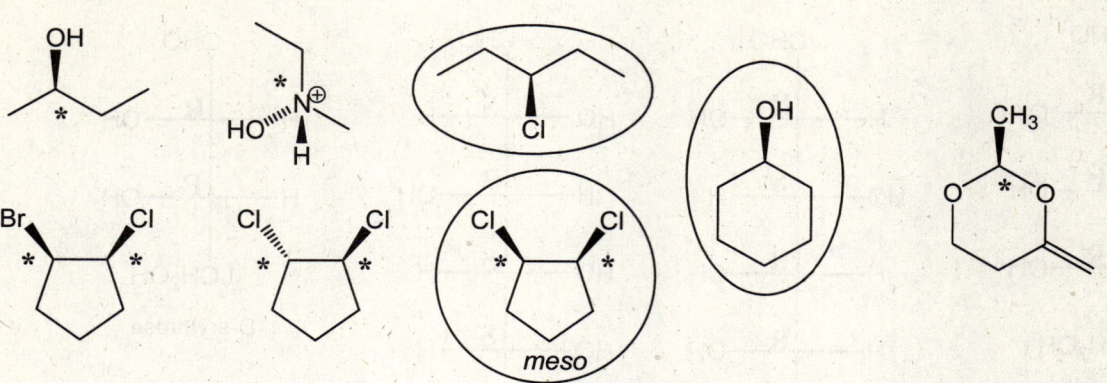

28.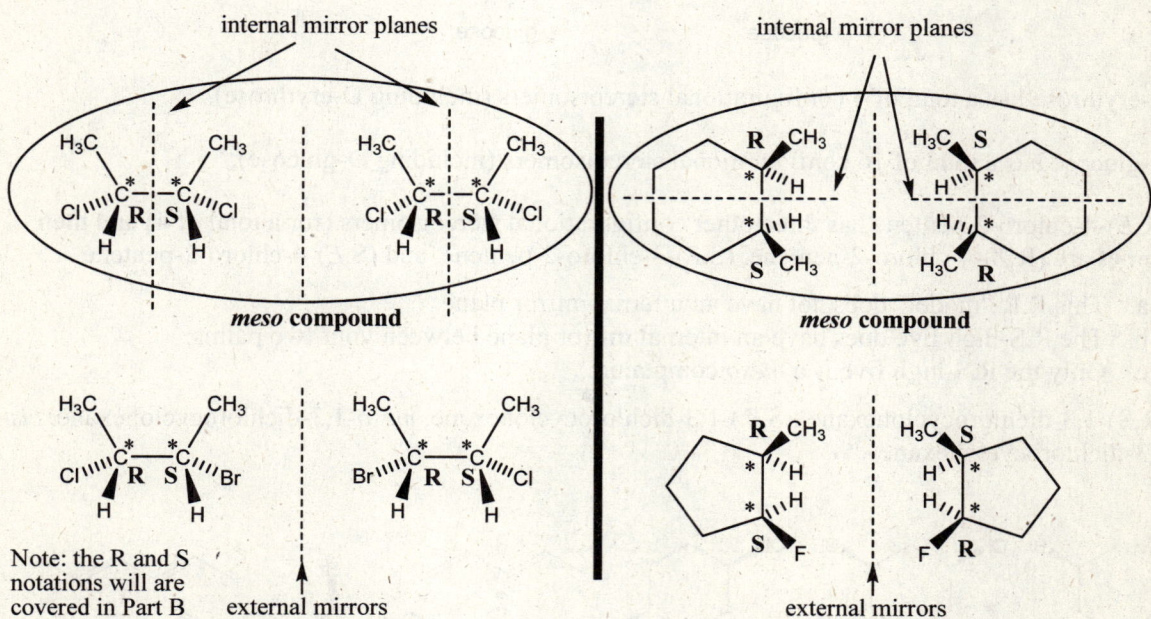

29. (–)-tartaric acid rotates plane-polarized **light** 12° counterclockwise ($\alpha_D = -12$).

30. a. The original solution was a racemic mixture, which will not rotate plane-polarized light.
 b. The term racemic mixture comes from *racemic acid,* the acid that Pasteur was analyzing that turned out to be a mixture of right and left handed molecules of the conjugate acid of tartrate. After Pasteur's discovery, the term racemic was used to describe any such solution with a 1:1 mixture of two enantiomers.
 c. Enantiomers

ChemActivity 9: Absolute Configuration (R/S)

Extend Your Understanding Questions (to do in or out of class)

16. *See below*

17. *See below.*

[Fischer projections shown: D-ribose (all R: H-R-OH, H-R-OH, H-R-OH with CHO top and CH₂OH bottom); D-glucose (R, S, R, R from top to bottom); L-glucose (S, R, S, S from top to bottom); D-erythrose (R, R)]

D-ribose   D-glucose   L-glucose   D-erythrose

18. D-erythrose has a total of 4 configurational stereoisomers (including D-erythrose).

19. D-glucose has a total of 16 configurational stereoisomers (including D-glucose).

20. (R,E)-4-chloro-2-pentene has three other configurational stereoisomers (for a total of 4) and their names are (R,Z)-4-chloro-2-pentene, (S,Z)-4-chloro-2-pentene, and (S,E)-4-chloro-2-pentene.

21. a. This R,R "model" does not have an internal mirror plane.
    b. The R,S-high five does have an internal mirror plane between your two palms.
    c. Only the R,S-high five is a *meso* compound.

22. (R,S)-1,3-dichlorocyclohexane, (S,R)-1,3-dichlorocyclohexane, *meso*-1,3-dichlorocyclohexane, *cis*-1,3-dichlorocyclohexane

23.

[Structures labeled 4, 3, 2, 2, 8, 3]

24.

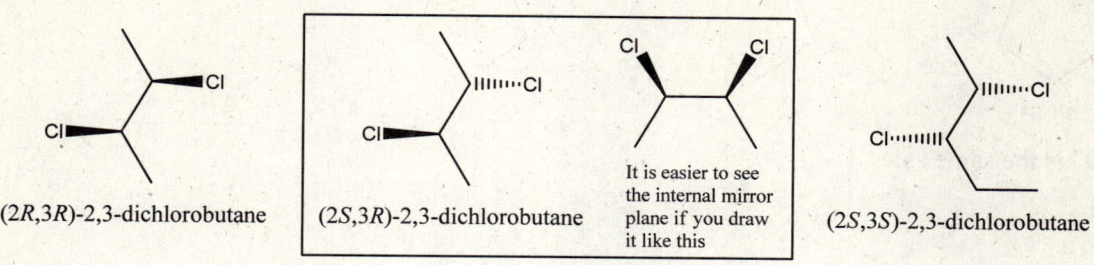

(2R,3R)-2,3-dichloropentane   (2S,3R)-2,3-dichloropentane   (2R,3S)-2,3-dichloropentane   (2S,3S)-2,3-dichloropentane

a.

(2R,3R)-2,3-dichlorobutane   (2S,3R)-2,3-dichlorobutane   [It is easier to see the internal mirror plane if you draw it like this]   (2S,3S)-2,3-dichlorobutane

b. 2,3-dichloropentane follows the 2^n rule, and so has four different stereoisomers. The butane molecule is symmetrical, which makes one of the possible stereoisomers a *meso* compound. This gives rise to redundancy. That is, the R,S stereoisomer is identical to the S,R stereoisomer, meaning there are only three unique possibilities.

25. The name *cis*-1,3-dichlorocyclohexane is explicit because there is only one way to draw this molecule (though there are four ways to name it, as shown in Exercise 10, above). The two chiral centers in *cis*-1,3-dichlorocyclohexane will be S and R, respectively (though you can list them in either order). The name *trans*-1,3-dichlorocyclohexane tells you that you must draw the two Cl groups on opposite sides of the ring plane, but it does not tell you if the two chiral centers should be both S, or both R. That is there are two kinds of *trans*-1,3-dichlorocyclohexane, (S,S) and (R,R).

Confirm Your Understanding Questions (to do at home)

26. When asked for a pair of configurational stereoisomers that are NOT enantiomers most students draw chiral structures such as the ones in Question 24 or 27. Students often forget that structures such as *cis* and *trans*-but-2-ene are also cofigurational stereoisomers of each other, and that they are also called diastereomers.

27. 1,3-dimethylcyclopentane configurational stereoisomers

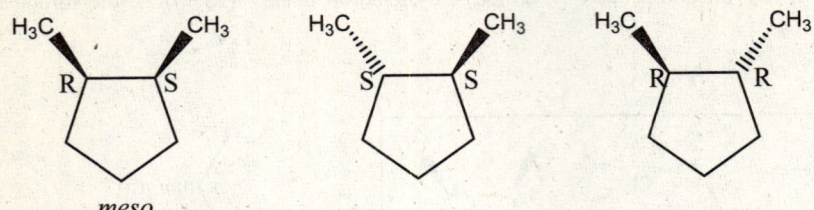

meso

28. "CHO" is the same as…

29. Regarding D-galactose and the reduced form of D-galactose…

    a. Because C=O is reduced to CH-OH (two H's are added to the double bond).

    b. Four in each

    c. Following the 2^n rule… $2^4 = 16$

    d. Some of the configurational stereoisomers of the reduced form will be *meso* compounds due to the greater symmetry of the reduced form (there is an OH group at both ends). This will lead to redundancy and a fewer number of total stereoisomers

    e. RRRR, RRRS, RRSS, RRSR, RSSS, RSSR, RSRS, RSRR
       SSSS, SSSR, SSRR, SSRS, SRRR, SRRS, SRSR, SRSS
       One algorithm for generating the above combinations is to recognize that there are 8 possibilities for three chiral centers (RRR, RRS, RSS, RSR, plus four others generated by switching R and S). Start by writing down both possibilities that start with RRR, that is, RRR(R) and RRR(S). Now write down both possibilities that start with RRS, that is, RRS(S) and RRS(R). Continue this by writing down both possibilities that start with RSS and RSR. This generates 8 stereoisomers (half of the 16). Now go back and do the same thing switching each R and S, to give the remaining 8 stereoisomers.

    f. RRRR, RRRS, RRSS, RRSR, RSSS, RSSR, RSRS,
       SSSS,   SRRS,   SRSS

       Note that the SSSR, SRRR, SSRS, RSRR, SSRR, and SRSR stereoisomers have been removed from the set of unique stereoisomers for reduced D-galactose because these are identical to the stereoisomers RSSS, RRRS, SRSS, RRSR, RRSS, and RSRS, respectively

30. For the following, refer to the molecule below...
    a. C₂ is attached to one C and 2 H's so holds a "hand" of CHH
       C₄ is attached to one C and 2 H's so holds a "hand" of CHH
       since this is a tie, we move on to examine the "hands" of C₁ (the largest group attached to C₂), and C₅ (the largest group attached to C₄).
    b. C₁ is attached to an O, and 2 H's so holds a "hand" of OHH
       C₅ is attached to 3 H's so holds a "hand" of HHH
    c. C₁'s hand beats C₅'s, so C₂ gets priority over C₄, which makes the chiral center S

31.

32. Both molecules are chiral.

Penicillin G   Alleve (Naproxen)

33. Penicillin G has 8 possible configurational stereoisomers; Aleve has 2.

34. When students are presented with this type of chart, there is a very strong temptation to memorize it. The value of this question is in drawing your own chart. Check your chart against those of your group mates. If you need further confirmation, there are many charts like this one in textbooks and on the internet.

ChemActivity 10: One-Step Substitution (S$_N$2)

Extend Your Understanding Questions (to do in or out of class)

21. Yes

22. A very good leaving group...

    a. has C—LG bond that is **easy** to break.

    b. is the conjugate base of a **strong** acid.

    c. is a very **weak** base.

    d. has a conjugate acid (H—LG) with a **low** pK_a

23. A **strong base nuclophile** (Nuc$^\ominus$)...

    a. in an S$_N$2 reaction will be downhill if the new Nuc—C bond is **strong**.

    b. will release a **large** amount of energy when the Nuc—C bond is formed.

    c. will be a **strong** base.

    d. Will have a conjugate acid (Nuc—H) with a **high** pK_a

24. a. *see previous question*
    b. *see below*
    c. iodide
    d. H$_2$S is a better nucleophile than H$_2$O; iodide is a better nucleophile than bromide; etc.

Memorization Task 10.3: Leaving Groups (table) and Nucleophiles (table)

R = H or alkyl

Very Good Leaving Groups	Good Leaving Groups	Poor Leaving Groups
RSO$_3^-$ (most often... R= CF$_3$, R= tol, or R= CH$_3$)	R$_2$O (water, alcohol, or ether)	F$^-$
I$^-$	Br$^-$	strong bases are very poor leaving groups
	Cl$^-$	RO$^-$
		R$_2$N$^-$
		R$_3$C$^-$

Table 10.1: Common Leaving Groups

Very Good Nucleophiles	Good Nucleophiles	Poor Nucleophiles
soft RS$^-$	Br$^-$	F$^-$
(niether) NC$^-$	R$_2$S	HCO$_3^-$
soft I$^-$	NR$_3$	R$_2$O
soft PR$_3$	Cl$^-$	(water, alcohol, or ether)
str. base *R$_3$C$^-$	RCO$_2^-$	
str. base *R$_2$N$^-$	N$_3^-$	All strong acids are very poor nucleophiles
str. base *RC≡C$^-$		
str. base *RO$^-$		

Table 10.2: Common Nucleophiles

tol = (toluene group) —CH$_3$

—S(=O)(=O)—tol is often abbreviated **Ts** (Tosyl group)

*These nucleophiles are also strong bases. They will NOT undergo S$_N$2 reactions when there is a faster acid-base pathway available (more on this later).

Confirm Your Understanding Questions (to do at home)

25. Reactant = R; Product shown in the question is also R, but the correct product (shown below) is S.

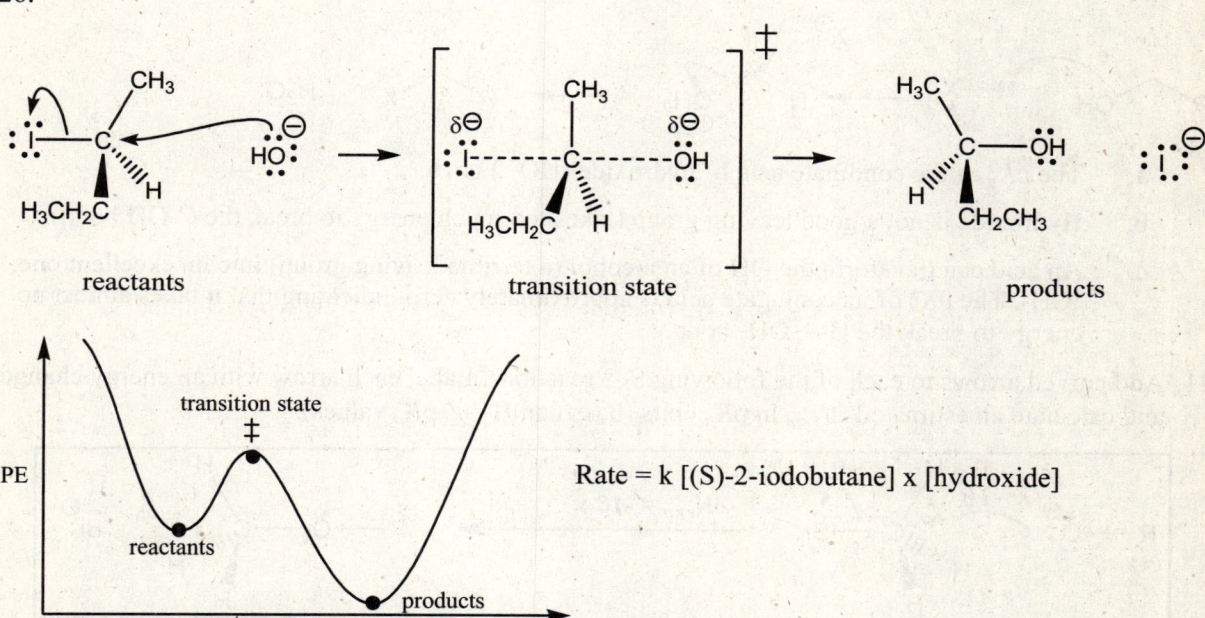

a. The correct product has an S absolute configuration because chiral inversion occurs during an S_N2 reaction.

b. *see above.*

26.

Rate = k [(S)-2-iodobutane] x [hydroxide]

27. An umbrella in the wind can sort of simulate this inversion because, if the umbrella is flexible enough, a strong wind can cause the supports on the umbrella to switch from being curved one way to being curved the other way. Most model sets are not flexible enough, and the holes are drilled on one side so they can't show this type of chiral inversion without switching two of the pieces.

28. The mechanism and neutral products of an S_N2 reaction between water and methyl iodide:

H_2O (water is a stronger base than iodide)

29. Use curved arrows to show the mechanisms of the following two acid-base reactions.

R—Ö—H ΔH = 3 → R—Ö(⊕)—H   :F:(⊖)
   -0 |
  H—F: H
   +3

:O: +5   H   ΔH = -4   :O: RNH₃(⊕)
 ‖   \\ | ‖
R—C—Ö—H   R—N—H   →   R—C—Ö:(⊖)
            |
            H
           -9

30. Mechanism of Synth. Transf. 10.1...

R—CH₂—OH   H—X   →   R—CH₂—OH₂(⊕)   X(⊖)   →   R—CH₂—X   H₂O

   a. The pK_a of the conjugate acid of hydroxide (HO⁻) is 16.

   b. Hydroxide is not a good leaving group (takes too much energy to break the C-OH bond).

   c. An acid can transform the OH of an alcohol (a terrible leaving group) into an excellent one: ⁺OH₂. The pK_a of its conjugate acid is approximately zero indicating that it takes almost no energy to break the H—⁺OH₂ bond.

31. Add curved arrows to each of the following S$_N$2 reactions. Label each arrow with an energy change, and calculate an estimated ΔH_{rxn} in pK_a units (based on H—Z pK_a values).

R—Ö:(⊖) -16 → H—C—Br: +0 → ΔH_{rxn} = -16 → R—Ö—C—H   :Br:(⊖)

H—Ö: -0 → H—C—Br: +0 → ΔH_{rxn} = 0 → H—Ö(⊕)—C—H   :Br:(⊖)
 | |
 H   H

:I:(⊖) -0 → H—C—OH +16 → ΔH_{rxn} = +16   **No Reaction** → :I—C—H   :ÖH(⊖)

:I:(⊖) -0 → H—C—OH₂(⊕) +0 → ΔH_{rxn} = 0 → :I—C—H   H₂Ö:

32. *See above.*

33. For each pair, circle the <u>better leaving group</u> (assume breakage of the bond highlighted in **bold**).

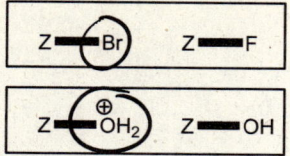 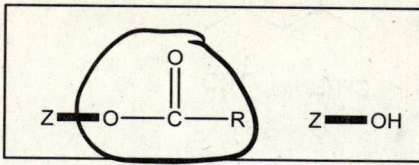 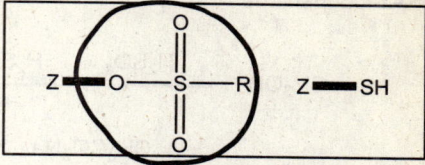

34. For the first four pairs, the better nucleophile is better based on the pK_a data cited. For the last two pairs, the circled nucleophile is a softer base (in keeping with the fact that the circled nucleophile is lower on the periodic table.

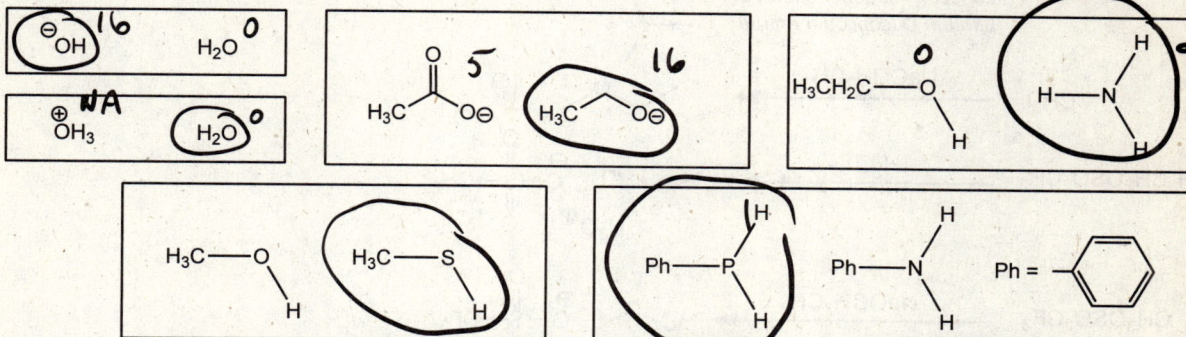

35.

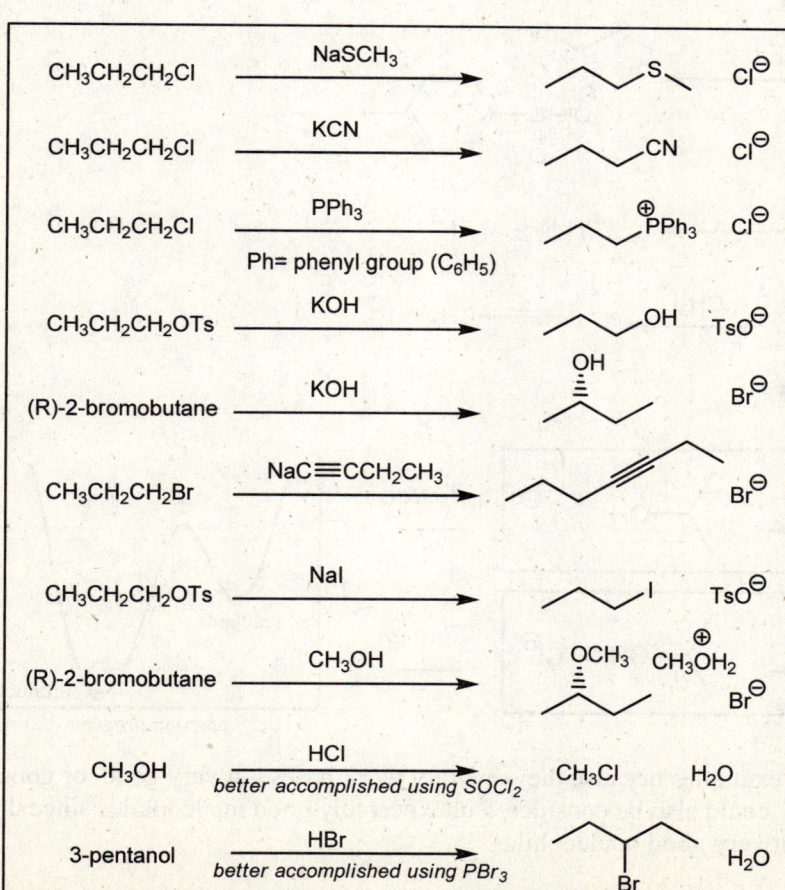

Continued on next page

(S)-2-iodopentane →[NaN₃] (S)-2-azidopentane (with N₃ group, wedge configuration)

CH_3OH →[H_2SO_4] →[H_2S] CH_3SH + H_2O

CH_3I →[CH_3CO_2Na] $CH_3CO_2CH_3$ (methyl acetate) + $I^{\ominus}$

CH_3I →[$((CH_3)_2CH)_2N^{\ominus}$ $Li^{\oplus}$, a special reagent called LDA, Lithium Diisopropyl Amide] $(iPr)_2N-CH_3$ + $I^{\ominus}$

CH_3I →[$NaOCH_2CH_3$] $CH_3CH_2OCH_3$ + $I^{\ominus}$

$CH_3CH_2OSO_2CH_3$ →[H_2O] CH_3CH_2OH + $^{\ominus}O-SO_2-CH_3$ + $H_3O^{\oplus}$

$CH_3OSO_2CF_3$ →[$NaOCH_2CH_3$] $CH_3OCH_2CH_3$ + $^{\ominus}O-SO_2-CF_3$

$CH_3CH_2CH_2OSO_2tol$ →[$NaBr$] $CH_3CH_2CH_2Br$ + $^{\ominus}O-SO_2-C_6H_4-CH_3$ (tosylate)

CH_3I → CH_3NH_2 + $I^{\ominus}$

$HCCH$ →[$NaNH_2$, NH_3] $HC\equiv C^{\ominus}$ →[CH_3I] $HC\equiv C-CH_3$ + $I^{\ominus}$

36. *see below*

[Box 1: $HO^{\ominus}$ attacks $(CH_3)_2CH-I$ → $(CH_3)_2CH-OH$ + $I^{\ominus}$] *Faster* - - - -

[Box 2: $HO^{\ominus}$ attacks $(CH_3)_2CH-Cl$ → $(CH_3)_2CH-OH$ + $Cl^{\ominus}$] *Slower* ——

[PE vs reaction progress diagram showing two curves: dashed (faster, lower transition state ‡) and solid (slower, higher transition state), both going from reactants to products]

37. I⁻, Br⁻, & Cl⁻ are the best examples because they are very weak bases but very good, or good nucleophiles. RS⁻ and NC⁻ could also be considered unexpectedly good nucleophiles since they are relatively weak bases but very good nucleophiles

38. **weak; low; strong; weak; good; down; good; large; soft; small; hard; soft**

39.

(CH₃)₂CH-ONa →[CH₃Br] (CH₃)₂CH-O-CH₃

CH₃CH₂CH₂-I →[NaCN] →[H₃O⁺] CH₃CH₂CH₂-COOH

NaC≡CCH₃ →[CH₃Br] →[Na, NH₃] CH₃CH=CHCH₃ →[Br₂/H₂O] CH₃CH(Br)CH(OH)CH₃

NaC≡CCH₃ →[CH₃Br] →[HgSO₄, H₃O⁺] CH₃C(O)CH₂CH₃

ChemActivity 11: Two-Step Substitution (S_N1)

Extend Your Understanding Questions (to do in or out of class)

11. Yes.

12. A two-step nucleophilic substitution reaction is called an S_N1 reaction because there is only one species involved in the slow step, and a one-step nucleophilic substitution is called an S_N2 reaction because there are two species involved in the slow step.

13.

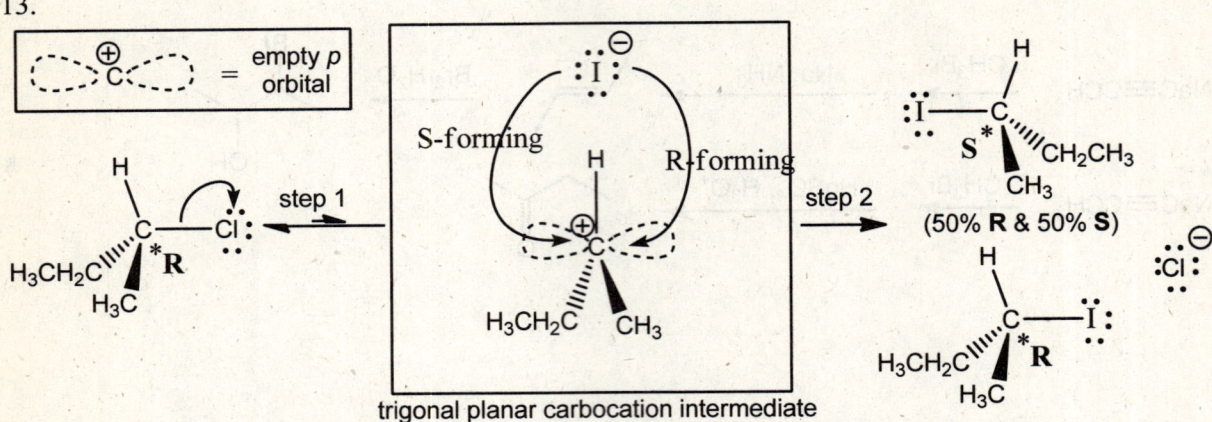

14. An S_N1 reaction at a chiral electrophilic carbon does NOT lead to chiral inversion the way an S_N2 reaction at a chiral center is known to do because the planar carbocation intermediate can be approached from either side by the nucleophile, giving an equal mixture of R and S products.

15. Racemic mixture.

16. Reaction 1 CANNOT form a favorable carbocation.

17. Reaction 3 will form the lowest potential energy (most favorable) carbocation when Br leaves.

18. Reaction 1 is least crowded and so MOST likely to proceed via a one-step (S_N2) mechanism.

19. Reaction 3 is too crowded to proceed via a one-step (S_N2) mechanism.

20. a. S_N2, S_N1, S_N2
    b. S_N1, S_N2, S_N1

21. When it comes to deciding whether a reaction will be S_N1 or S_N2, **electronic factors** (carbocation stability) and **steric factors** are generally in agreement. That is, you almost always get the same answer (e.g. Reaction 1 in Model 4 goes by S_N2) if you consider carbocation stability or if you consider steric factors.

22. According to Model 5, a **polar protic** solvent is best for an S_N1 reaction.

23. A polar protic solvent is best for an S_N1 reaction because the intermediate of an S_N1 (and therefore by the Hammond postulate, the transition state) is very ionic. [The intermediate *is* an ion.] Since a polar protic solvent is as a super-polar solvent, and the more polar a solvent the better it is at dissolving and stabilizing ions, it follows that a polar protic solvent will lower the energy of the transition state and intermediate of an S_N1, thereby speeding the reaction.

24. Yes, the energy diagram below, right shows that the dotted line (in polar protic solvent) has a lower energy transition state and intermediate than the reaction in a less polar solvent (solid line).

25. An S_N2 reaction is faster in polar (not polar protic) solvent. *See arrows on diagram below, left.*

26. The answer to Question 23 explains why a polar protic solvent speeds an S_N1 reaction (by stabilizing the carbocation intermediate), but it also impedes (slows) an S_N2 reaction. This is because the polar protic solvent lowers the potential energy of the nucleophile in the S_N2 reaction, thereby increasing the activation energy and slowing the reaction. The reason the polar protic solvent lowers the potential energy of the starting material in most S_N2 reactions is that in most cases the nucleophile is an ion. By stabilizing this ion, the solvent deactivates the nucleophile. Another way of thinking about this is that if you are relying on an ionic nucleophile to drive a reaction, a solvent that can hydrogen bond to this nucleophile and thereby compete with the electrophile, slowing the S_N2 reaction.

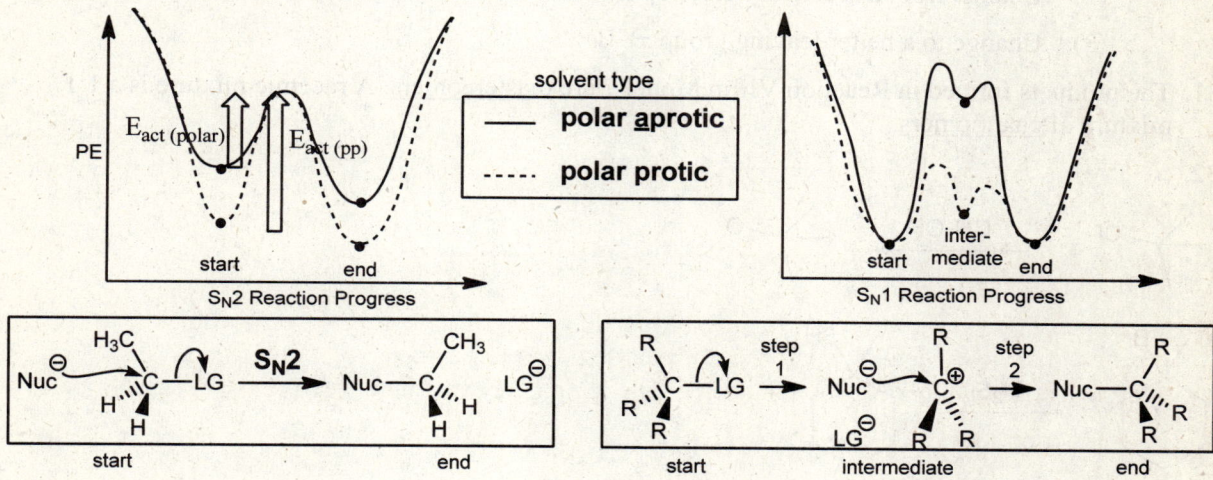

Confirm Your Understanding Questions (to do at home)

27. Only hydroxide is a good nucleophile because it is small enough to form a bond to the electrophilic carbon. The large size of *tert*-butoxide hinders it from making this bond.

28.   a.   Rxn A is slower than Rxn B because chloride is a worse leaving group than iodide.

      b.   Rxn A is slower than Rxn C because chloride is a worse nucleophile than hydroxide.

      c.   Rxn A is slower than Rxn D because a 2° electrophilic carbon is more hindered than a 1° electrophilic carbon.

29. The extreme steric hindrance of the S_N2 transition state explains its high energy, and the fact that the S_N1 pathway is much faster. (a, b: *see below*)

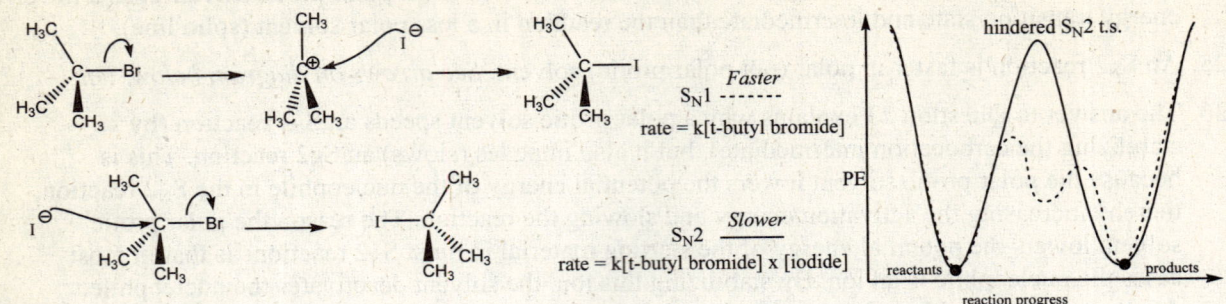

S_N1 ------ *Faster*
rate = k[t-butyl bromide]

S_N2 ——— *Slower*
rate = k[t-butyl bromide] x [iodide]

30.   (1) Change to a nucleophile that is better in that it is a *stronger base* → A

    (2) Change from a secondary electrophilic C to a primary electrophilic carbon → C

    (3) Change to a better leaving group → B

31. The products formed in Reaction VII in Model 7 are <u>diastereomers</u>. A racemic mixture is a 1:1 mixture of <u>enantiomers</u>.

32.

we will learn in the next ChemActivity that these last two reactions will likely yield alkenes instead of the product shown

33. These primary alkyl halides form resonance stabilized (allylic or benzylic) carbocations. This allows them to undergo S_N1.

34. These allylic and benzylic alkyl halides CANNOT undergo an S_N2 regardless of solvent because tertiary alkyl halides are too hindered to undergo S_N2.

35. A polar protic solvent speeds the rate of an S_N1 mechanism by lowering the potential energy of the transition state, and thus lowering the activation energy. Chemists often talk about polar protic solvents lowering the energy of the intermediate carbocation. Technically, lowering the PE of the intermediate would not (alone) speed the reaction; however, according to the Hammond postulate, lowering the energy of the intermediate is likely to also lower the PE of the transition state.

36. A polar protic solvent <u>slows</u> the rate of an S_N2 mechanism because it dramatically lowers the PE of the starting material, but does not have as large a stabilizing effect on the transition state. The net result is that the activation barrier is increased, and the rate is slowed. One way to think about this is that a polar protic solvent hydrogen bonds to the nucleophile, thus deactivating it, and simply getting in the way of making a new bond to the electrophilic carbon.

37. *Answer continued on the next page.*

Volume 1, ChemActivity 11: Two-Step Nucleophilic Substitution (S$_N$1)

Substrate	Reagent	Products		Mechanism
(S)-3-deutero-1-iodopentane	NaN$_3$	N$_3$–CH$_2$CH$_2$–C(S)(D)(H)–CH$_2$CH$_3$	I$^-$	S$_N$2
CH$_3$OH	H$_2$SO$_4$, H$_2$S	CH$_3$SH, H$_2$O		acid-base followed by S$_N$2
CH$_3$I	CH$_3$CO$_2$Na	CH$_3$–O–C(=O)–CH$_3$	I$^-$	S$_N$2
CH$_3$OH	((CH$_3$)$_2$CH)$_2$N$^-$ Li$^+$	((CH$_3$)$_2$CH)$_2$NH	CH$_3$O$^-$	acid-base
cyclohexyl–OSO$_2$CH$_3$	HOCH$_3$	cyclohexyl–OCH$_3$	$^-$O–SO$_2$–CH$_3$	S$_N$1
CH$_3$CH$_2$OSO$_2$CH$_3$	H$_2$O	HOCH$_2$CH$_3$	$^-$O–SO$_2$–CH$_3$	S$_N$2
F$_3$C–C(=O)OH	NaOCH$_2$CH$_3$	F$_3$C–C(=O)O$^-$	HOCH$_2$CH$_3$	acid-base
(CH$_3$)$_3$C–OTs	NaBr	(CH$_3$)$_3$C–Br	$^-$OTs	S$_N$1
CH$_3$I	NaN(CH$_3$)$_2$	(CH$_3$)$_3$N	I$^-$ (–N$^+$(CH$_3$)$_4$ also likely)	S$_N$2
(H$_3$C)$_3$CO–SO$_2$–C$_6$H$_4$–CH$_3$	HOCH$_3$	(H$_3$C)$_3$C–OCH$_3$	$^-$OTs	S$_N$1

38.

PhCH(Br)CH$_3$, H$_2$O — more likely to undergo S$_N$1

PhCH$_2$CH$_2$Br, HS$^-$ — more likely to undergo S$_N$2

ChemActivity 12: Two-Step Elimination (E1)

Extend Your Understanding Questions (to do in or out of class)

13.

14.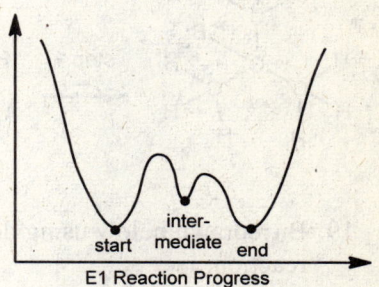

  a. The reaction rate is dependent on the concentration of the starting material (R—Br), but not on concentration of the base (bicarbonate).

  b. Rate expression for this two-step elimination:
  rate = k [RBr]

  c. A two-step elimination is called an "**E1**" reaction. The "E" stands for elimination, and the "1" signifies the number of species involved in the slow step.

  d. The second step in an E1 reaction is rate-limiting.

Confirm Your Understanding Questions (to do at home)

15.

16.

17. See answers to Exercise 15, and energy diagram below.

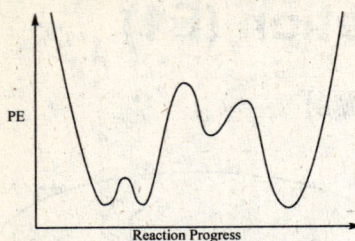

18. Br⁻ does react with the carbocation, but the product is simply the starting alkyl halide.

19. B (redrawn below using dotted lines for partial bonds) is the transition state of step 2 of an E1 reaction.

20.

21. Ethene is the only unsubstituted alkene. It is called unsubstituted because there are no alkyl groups attached to the carbons involved in the double bond.

22. The more substituted a double bond, the lower its PE.

23.

24. This is an S_N1 reaction.

25.

26.

27.

ChemActivity 13: One-Step Elimination (E2)

Extend Your Understanding Questions (to do in or out of class)

13.

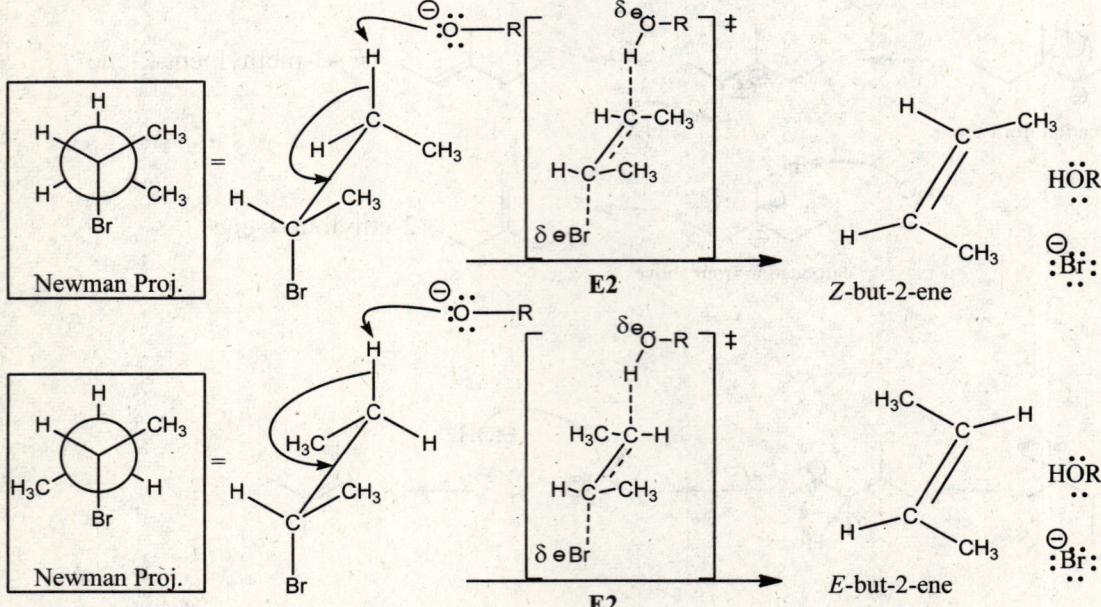

14. The answer to the previous question is consistent with Mem. Task 12.2 from ChemActivity 12.

15.

Lower PE, will spend more time in this conformation

16. Below are Newman and "sawhorse" representations of the two conformations of (S)-2-bromobutane that can lead to E2 reactions, along with the products of these E2 reactions.

17. *gauche, Z, anti, E*

18. The ***E*-2-butene** product will be more prevalent in the product mixture because the *anti* conformation leading to this product is more favorable, and therefore represents a larger proportion of the reaction mixture at any instant.

19. This fits with the answer to Question 15b (that the *anti* conformation is more favorable/prevalent than the *gauche* conformation.

20. The answer to the previous two questions is consistent with the this product distribution because the *trans* stereoisomer dominates, and 1-butene is minor among the alkene products.

21.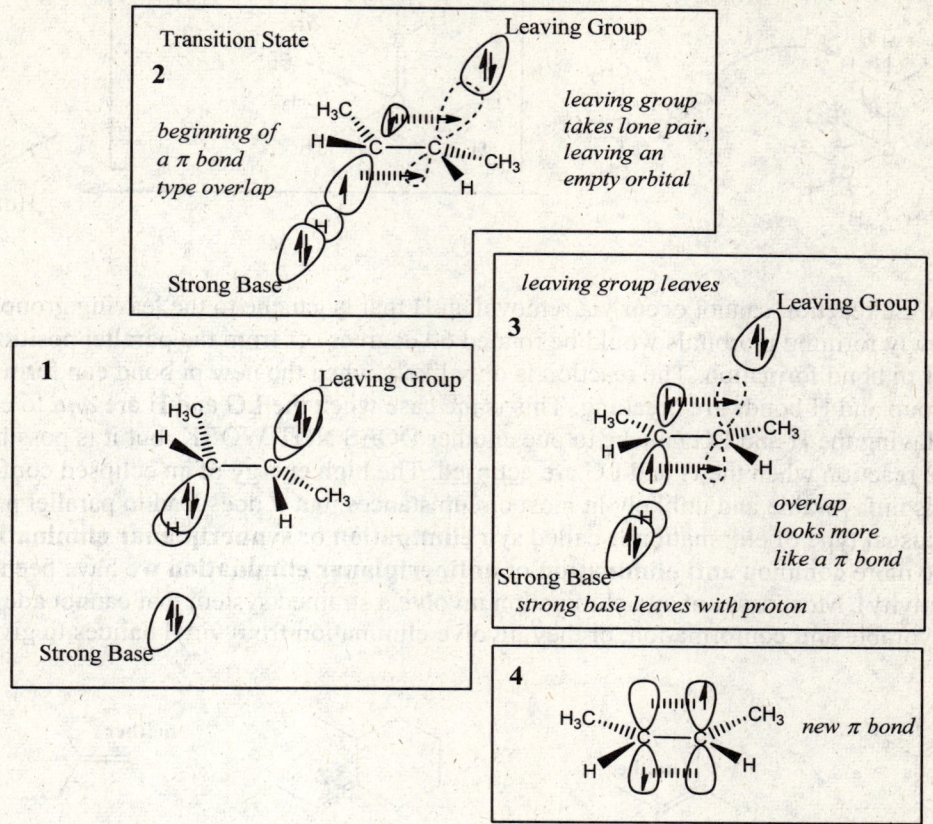

Confirm Your Understanding Questions (to do at home)

22.

23.   a.   A weak base is not strong enough to initiate proton removal as is required in E2.

   b.   A strong base *can* undergo E1 only if E2 is impossible (e.g. there is no H *anti* to the leaving group). This generalization holds true because, with a strong base, E2 is much faster than E1.

24. The major product of this reaction is the Hofmann product.

25. An E2 reaction cannot occur via removal an H that is gauche to the leaving group because the newly forming p orbitals would be rotated 60 degrees off from the parallel position that is required for pi bond formation. The reaction is only likely when the new pi bond can form as the leaving group and H bonds are breaking. This is the case when the LG and H are *anti* to each other. [Having the H and LG gauche to one another DOES NOT WORK, but it is possible to undergo an E2 reaction when the H and LG are eclipsed. The high energy of an eclipsed conformation makes this unfavorable and unlikely in most circumstances, but it does lead to parallel p orbitals. This unusual type of elimination is called **syn elimination** or **synperiplanar elimination** (in contrast to the more common **anti elimination** or **antiperiplanar elimination** we have been studying in this activity). Most cases of syn elimination involve a strained system that cannot adopt the more favorable anti conformation, or they involve elimination from vinyl halides to give alkynes.]

26.

27.

28.

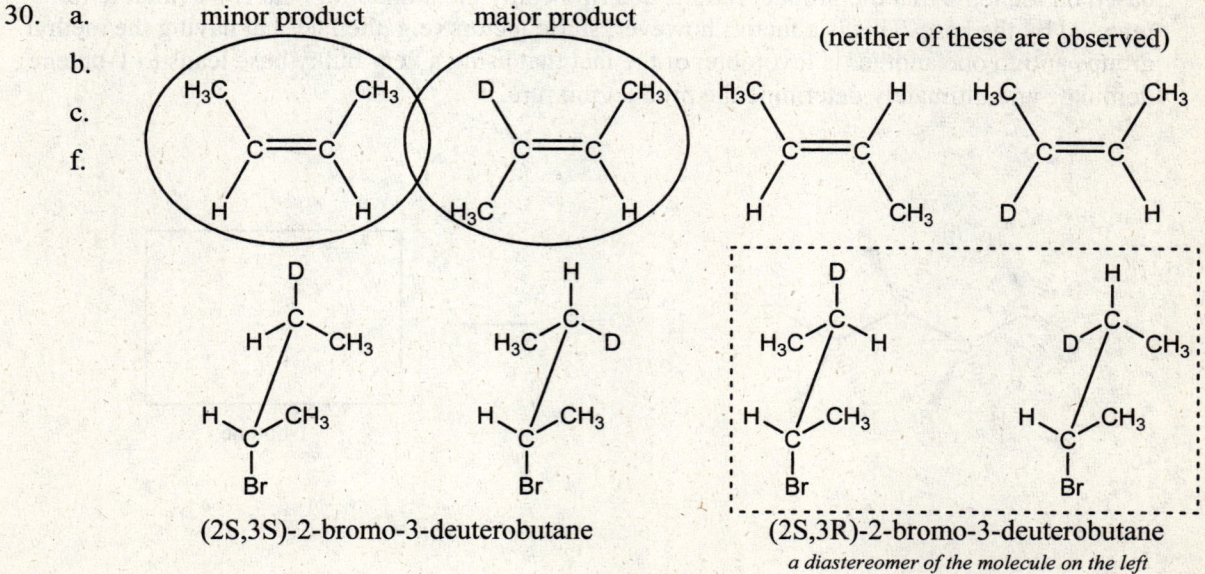

29. (3S,4S)-3-bromo-4-deuterohexane. Its enantiomer is called (3R,4R)-3-bromo-4-deuterohexane.

a. (3R,4R)-3-bromo-4-deuterohexane

b.

c.

d. (3R,4S)-3-bromo-4-deuterohexane

e. The molecule in c. is a diastereomer of the original starting material. (It is also a diastereomer of the molecule in part a. of this question.

30. a.
b.
c.
f.

minor product   major product   (neither of these are observed)

(2S,3S)-2-bromo-3-deuterobutane   (2S,3R)-2-bromo-3-deuterobutane
a diastereomer of the molecule on the left

Continued on next page.

d. (2R,3R)-2-bromo-3-deuterobutane

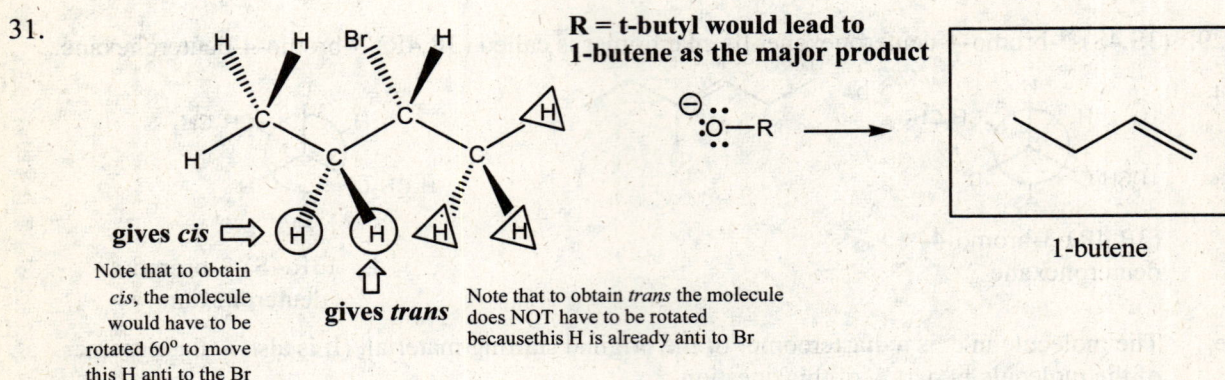

e. No, (2R,3R)-2-bromo-3-deuterobutane cannot give rise to the two uncircled alkenes above.

g. Only a diastereomer of (2R,3R) or (2S,3S) can give rise to the uncircled products. Two sawhorse representations of one of the two possible diastereomers (2S,3R) is shown above.

31.

g. The prediction that 1-butene is 3 times more likely to form than *trans*-2-butene appears to be based on the logic that the product ratio is determined by the number of beta H's available for removal by the base. This is a factor; however, steric factors (e.g. the fact that having the methyl groups anti to one another is favorable; or the fact that using a very bulky base leads to 1-butene) dominate and ultimately determine the product mixture.

ChemActivity 14: S_N1, S_N2, E1, E2 in Competition

Extend Your Understanding Questions (to do in or out of class)

10.

Favored Mech.	Base/Nuc	R—X	Rate dependent on [??]	Solvent	Temp
S_N1	Weak base and ok nucleophile	2°, 3°, allylic or benzylic only	R–X only	polar-protic required	cool
S_N2	Good nucleophile	methyl and 1° best but 2° ok	R–X and Nucleophile	polar aprotic	cool
E1	Weak base and poor nucleophile	2°, 3°, allylic or benzylic only AND must have H_β	R–X only	polar-protic required	hot
E2	Strong base	1°, 2°, 3° allyl and benzyl all ok but must have H_β	R–X and Base	polar aprotic	hot

11. Only S_N2 (no E2) is observed in Rxn A because E2 requires a beta hydrogen.

12. Only E2 (no S_N2) is observed in Reaction B because a tertiary electrophilic carbon is too sterically hindered to undergo an S_N2 reaction. (But there are plenty of beta hydrogens.)

13.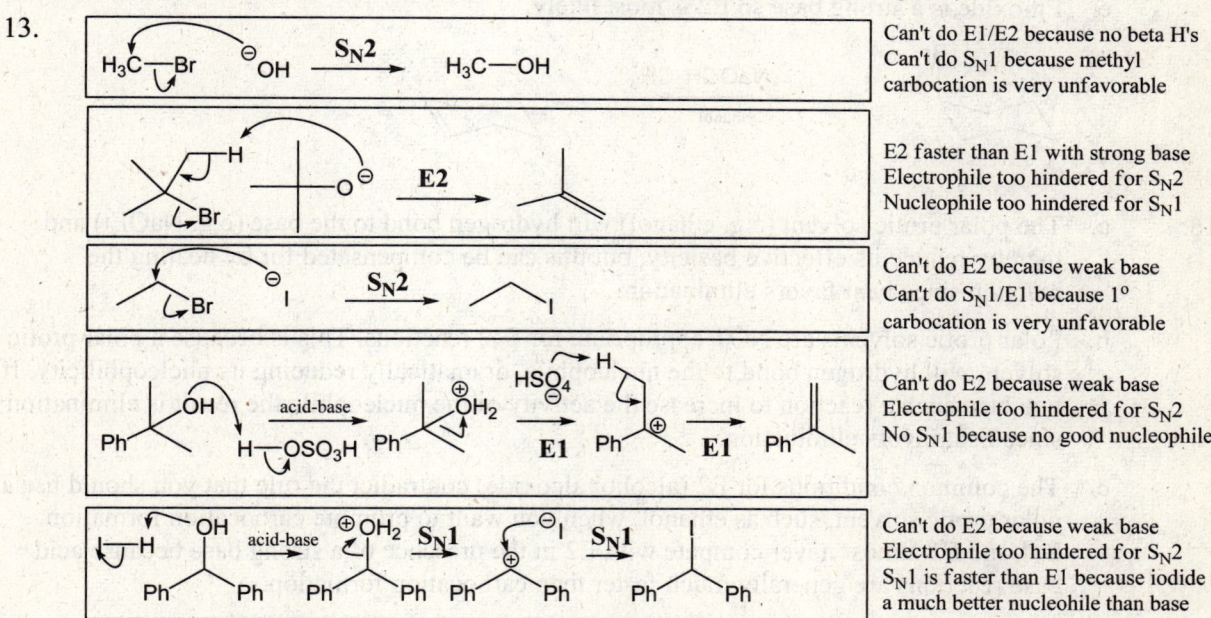

Can't do E1/E2 because no beta H's
Can't do S_N1 because methyl carbocation is very unfavorable

E2 faster than E1 with strong base
Electrophile too hindered for S_N2
Nucleophile too hindered for S_N1

Can't do E2 because weak base
Can't do S_N1/E1 because 1° carbocation is very unfavorable

Can't do E2 because weak base
Electrophile too hindered for S_N2
No S_N1 because no good nucleophiles

Can't do E2 because weak base
Electrophile too hindered for S_N2
S_N1 is faster than E1 because iodide is a much better nucleohile than base

14. a. ***S_N1 and E1 cannot occur with a 1° leaving group***. This statement is TRUE (unless you are talking about a primary allylic or primary benzylic leaving group). Both of these mechanism require a carbocation, and ordinary 1° carbocations are unfavorable.

   b. ***S_N2 and E2 cannot occur with a 3° leaving group***. This statement is FALSE, and reflects a very common misconception. It is true that S_N2 cannot occur at a 3° electrophilic carbon, but E2 does not share this restriction. (See the Exercise 22, above, for an example.)

15. a. This reaction is most likely to undergo S_N1, though a mixture of two different S_N1 products is likely. If the reaction were heated, E1 may also take place.

b. In neutral solution the OH group is not a good leaving group, so reaction can only occur at the primary alkyl halide position. Cyanide is a good nucleophile so S_N2 is the only possibility.

c. Methoxide is a strong base so E2 is most likely. Since it is a small base, the major product is expected to follow Zaitsev's rule.

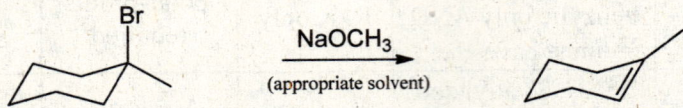

d. In neutral solution the OH group is not a good leaving group so no reaction can take place.

e. Ethoxide is a strong base so E2 is most likely.

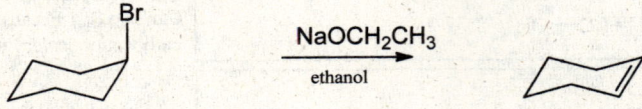

16. a. The polar protic solvent (e.g. ethanol) will hydrogen bond to the base (e.g. NaOEt) and thereby reduce its effective basicity, but this can be compensated for by heating the reaction since heat favors elimination.

b. Polar protic solvents are NOT appropriate for S_N2 reactions. This is because a polar protic solvent will hydrogen bond to the nucleophile, dramatically reducing its nucleophilicity. If you heat such a reaction to increase the activity of the nucleophile the result is elimination since heat favors elimination.

c. The common conditions for E2 (alcohol/alcoxide) contradict the rule that you should use a polar protic solvent, such as ethanol, when you want to promote carbocation formation. S_N1 and E1 almost never compete with E2 in the presence of a strong base because acid base reactions are generally much faster than carbocation formation.

17. *For one set of examples, see answers to Question 12.*

ChemActivity 15: Retrosynthesis

Extend Your Understanding Questions (to do in or out of class)

20. Cyanide (NC⁻, alkynyl anions RC≡C⁻, and R₃C⁻
    a. The other two carbon nucleophiles from the table both have a terminal triple bond.
    b. One carbon is need to go from starting material to target.

21.
   a. styrene —HBr/peroxides→ —NaCN→ —acid/water→ 3-phenylpropanoic acid (target)

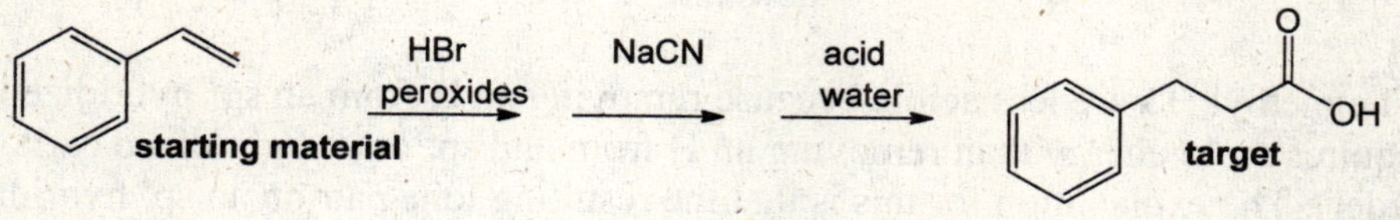

   b. styrene —HBr/peroxides→ —HC≡CNa→ —H₂, Lindlar Catalyst→ —1) KMnO₄ (hot) 2) acid/water→ 3-phenylpropanoic acid (target)

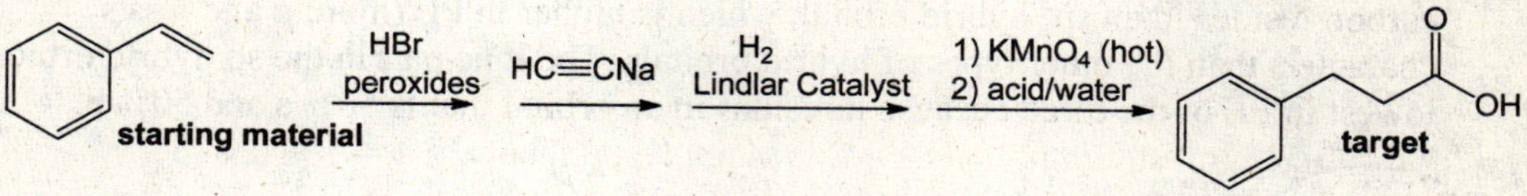

22. Below are examples of molecules containing various functional groups.

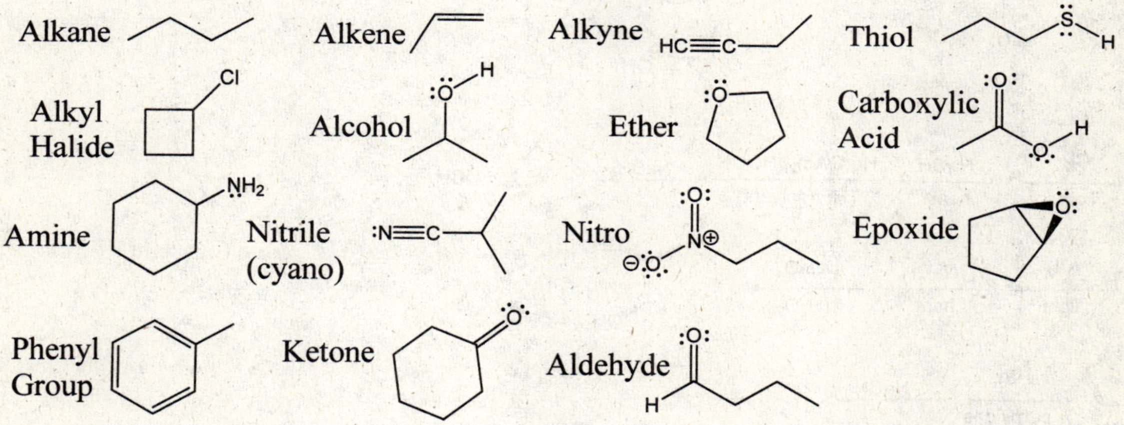

- Alkane
- Alkene
- Alkyne (HC≡C-)
- Thiol
- Alkyl Halide
- Alcohol
- Ether
- Carboxylic Acid
- Amine
- Nitrile (cyano) (:N≡C-)
- Nitro
- Epoxide
- Phenyl Group
- Ketone
- Aldehyde

Confirm Your Understanding Questions (to do at home)

23.
a) 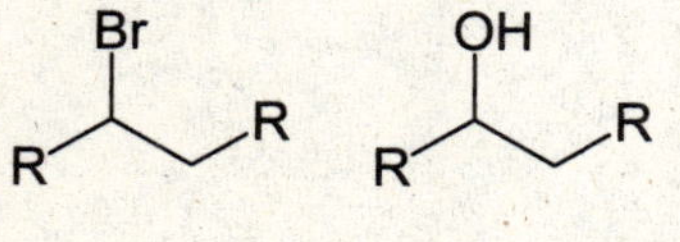 R-CHBr-CH₂-R ; R-CH(OH)-CH₂-R
b) 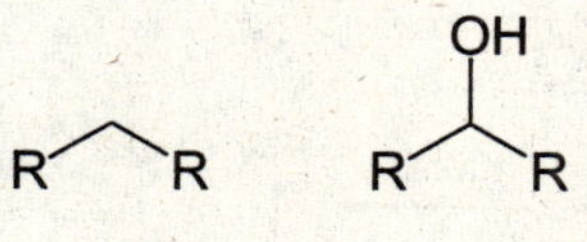 R-CH₂-R ; R-CH(OH)-R
c) 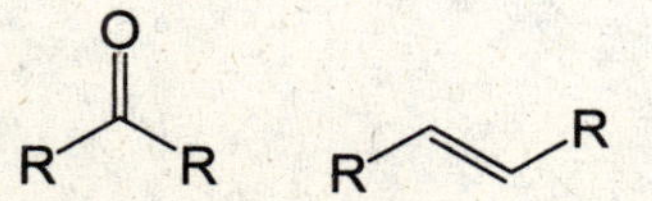 R-CO-R ; R-CH=CH-R (cis/trans)
d) 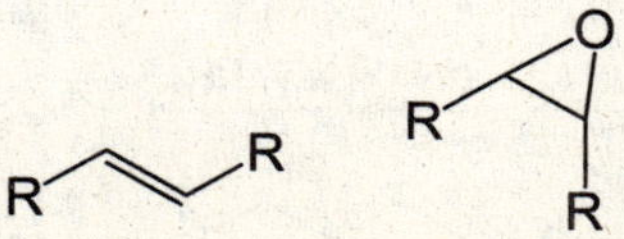 R-CH=CH-R ; R-epoxide-R

24.

[Top reaction: propene + :NH₂⁻ (H) → vinyl anion + NH₃]
+45 / −35, ΔH = +10 pK_a units, uphill

[Bottom reaction: H₃C—C≡C—H + :NH₂⁻ (H) → H₃C—C≡C:⁻ + NH₃]
+25 / −35, ΔH = −10 pK_a units, downhill

e. The methyl H's are less acidic because removing an H from an sp^3 hybridized carbon requires more energy than removing an H from and sp^2 or sp hybridized carbon (in that order). The explanation for this is that the resulting lone pair on an sp^3 hybridized carbon resides in an sp^3 hybrid orbital, which is higher in PE (more p and less s character) than the other types of hybrid orbitals. The lone pair in the sp hybrid orbital is lowest in PE of the three because it resides in an orbital that is 50% s and 50% p.

25. [Mechanism: ⁻OH + H₃C—CHCl—CHCl—H → H₃C—CH=CHCl + ⁻OH → H₃C—C≡CH]

26.

cyclopentane →(Cl₂, light)→ →(NaOH, heat)→ →(Hg(OAc)₂, HOCH₃, THF (solvent))→ →(NaBH₄)→ cyclopentyl—OCH₃

cyclohexane →(Cl₂, light)→ →(NaOH, heat)→ →(1) OsO₄ 2) H₂O₂)→ trans-cyclohexane-1,2-diol (OH, OH)

vinylcyclopentane →(HBr, peroxides)→ →(NaCN)→ cyclopentyl—CH₂CH₂—CN

ethylbenzene →(NBS)→ →(EtOH, heat)→ →(Br₂, water)→ Ph—CH(OH)—CH₂Br

propene →(NBS)→ →(NaOH)→ →(KMnO₄)→ acetone

cyclohexene →(Hg(OAc)₂, H₂O, THF (solvent))→ →(NaBH₄)→ →(KMnO₄)→ cyclohexanone

HC≡CH →(NaNH₂)→ →(CH₃I)→ →(NaNH₂)→ →(CH₃I)→ →(Na/NH₃)→ trans-2-butene

cyclohexyl—CH₂Br →(NaOH, heat)→ →(Br₂, water)→ 1-(bromomethyl)cyclohexan-1-ol

27.

[Reaction schemes showing multi-step synthesis sequences:]

- Cyclopentyl → NBS → (CH₃)₃CONa, heat → HBr, peroxides → NaCN → H₃O⁺ → cyclopentyl-CH₂-CO₂H

- Butane → NBS → (CH₃)₃CONa, heat → 1) BH₃ THF, 2) H₂O₂, NaOH water → Na → propyl-I → dipropyl ether

- Pentene → HBr, peroxides → NaC≡CCH₃ → Na/NH₃ → trans-alkene

- Isopropylbenzene (cumene) → NBS → EtOH, heat → 1) O₃, 2) Zn, HOAc → NaC≡CH → H₃O⁺, heat → H₂, Pt → sec-butylbenzene

- 2-pentanol (OH) → PBr₃ → (CH₃)₃CONa, heat → 1), 2) → butanal

- Methylcyclohexane → NBS → (CH₃)₃CONa, heat → HBr, peroxides → NaC≡CH → H₂O, H₂SO₄, HgSO₄ → cyclohexyl methyl ketone

- Cyclopentene → Br₂ → NaOH, heat → 1) O₃, 2) Zn, HOAc → methylmalondialdehyde

- Cyclohexane → NBS → (CH₃)₃CONa, heat → MCPBA → PrOH, H₂SO₄ → cyclohexyl propyl ether with OH

28. An electron pair in an *sp* hybrid orbital is lower in potential energy than an electron pair in an sp^2 hybrid orbital because the latter resides in an sp^2 hybrid orbital, which is higher in PE (more p and less s character) than an sp hybrid orbitals. That is, the lone pair in an sp hybrid orbital is lower in PE because it resides in an orbital that is 50% s and 50% p.

29. It takes less energy to convert ethyne into its conjugate base because of the argument in the previous question.

30. The C—H bond in ethyne is shorter than the C—H bond in ethene or ethane (even though it is easier to break with a base) because the ethyne C—H bond is made from overlap of an sp orbital and an s orbital, whereas the ethene C—H bond is made from overlap of an sp^2 hybrid orbital and an s orbital. The sp orbital is closer to the carbon, thus it pulls the H closer to the carbon than in ethene (or ethane).

31. True. The fact that it is possible to selectively reduce the first pi bond implies that the second one is less reactive. Also, a Lindlar catalyst is an attenuated palladium catalyst, often called "poisoned". The additives reduce the reactivity of the Pd so that only the first pi bond can react.

ChemActivity 16: Carbon (^{13}C) NMR

Extend Your Understanding Questions (to do in or out of class)

13. *See below*

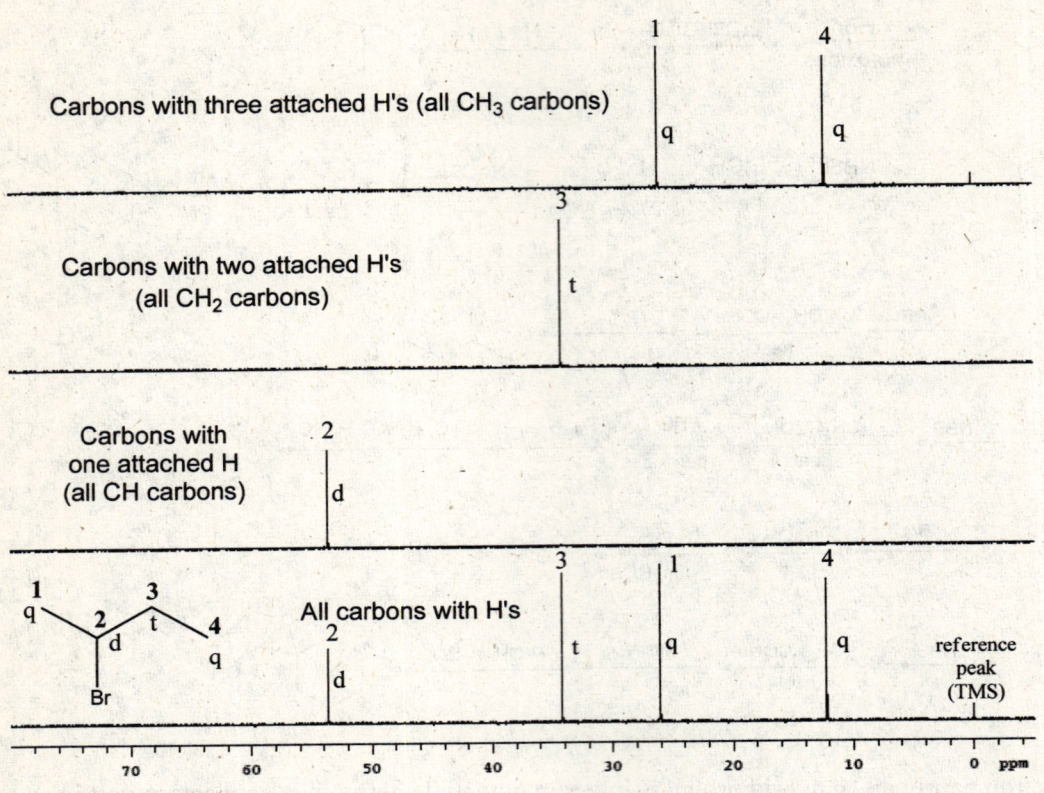

Figure 16.4: DEPT NMR Spectrum of 2-bromobutane

14. The TMS peak appears on the bottom and top levels of the DEPT spectrum because TMS contains four equivalent CH_3 groups.

15. a. *see next page*; b. Carbon 2 does not show up on the DEPT;
    c. The peak for carbon 2 will appear around 200 ppm on a regular ^{13}C NMR spectrum.

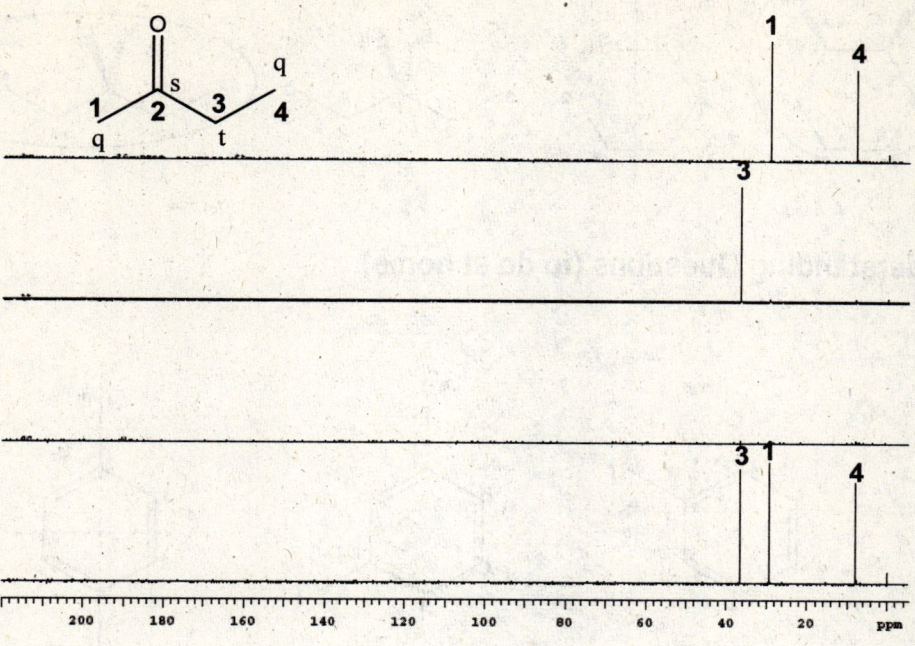

16. a. *see below*;   b. *see below*;   c. Object 3 has no additional mirror planes.

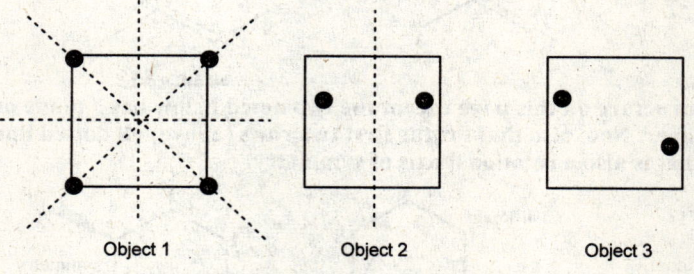

17.

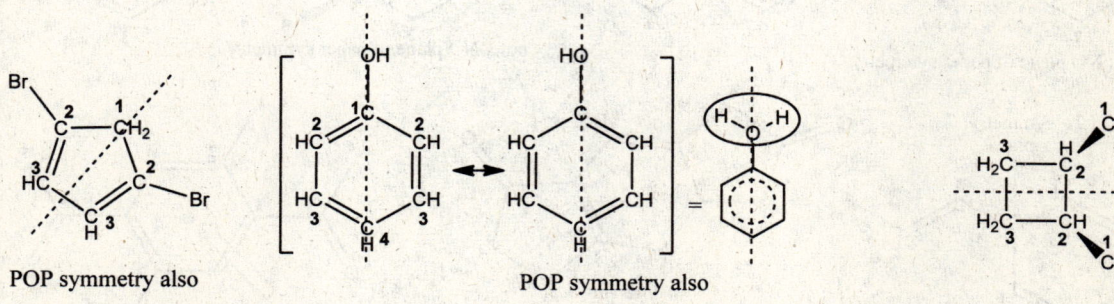

18. *Confirm with models.*

19. *See above.*

20. For the following molecules, the plane of the paper (POP) is NOT a symmetry plane.

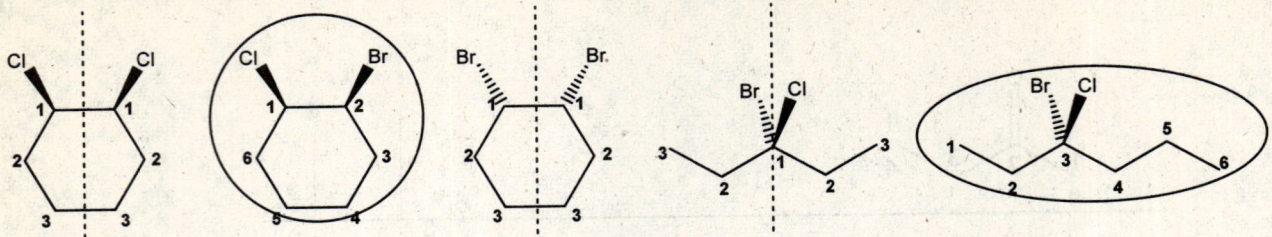

Confirm Your Understanding Questions (to do at home)

21.

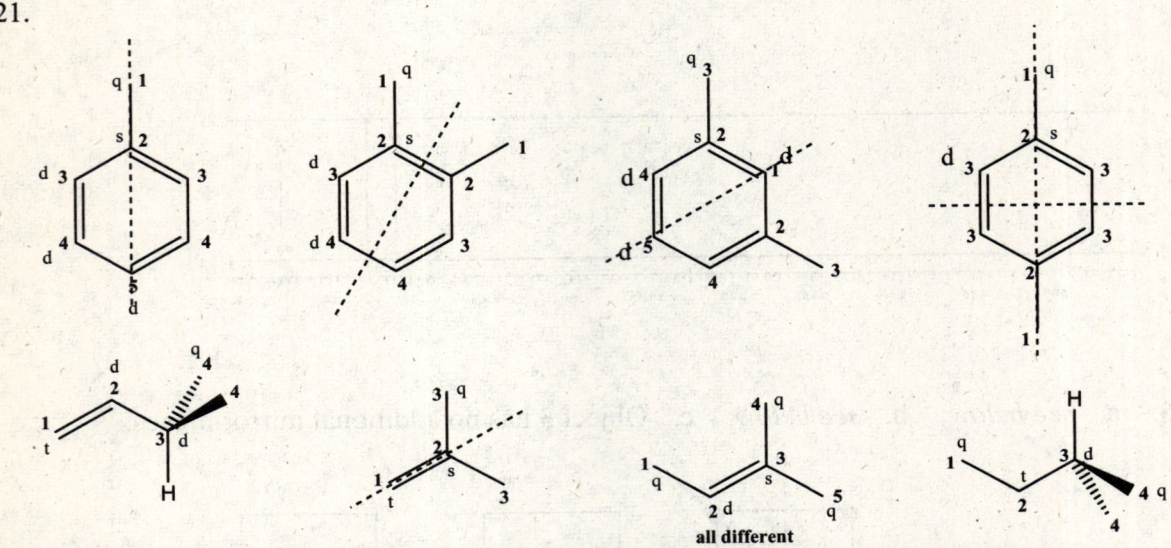

As drawn, every structure on this page except the two noted below has a plane of symmetry in the plane of the paper. Note also that for the first two rows (above) all dotted lines represent a symmetry plane that is also a rotational axis of symmetry.

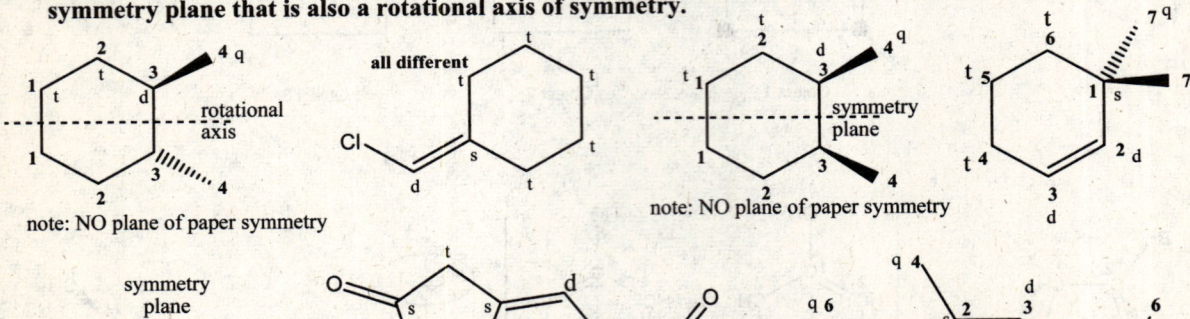

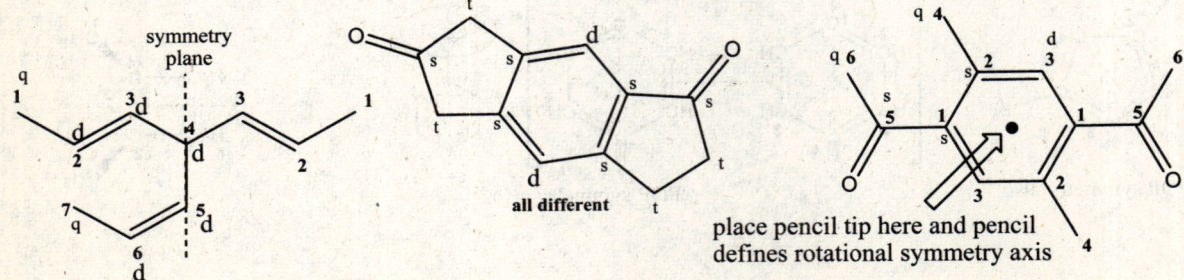

22. Complete the sentence: In coupled ^{13}C NMR, the number of peaks in a peak cluster ("multiplicity of the peak cluster") tells you the number of hydrogens attached to that C.

23. Complete the sentence: In ^{13}C NMR, the location of the peak cluster along the x axis (ppm value) tells you something about the hybridization and/or its proximity to functional groups such as OH, Cl, etc.

24. Complete the sentence: In decoupled ^{13}C NMR, the number of peaks tells you the number of unique (chemically distinct) carbons.

25. Complete the sentence: In decoupled ^{13}C NMR, the height of a peak tells you almost nothing.

26. Chemists rarely use proton-coupled ^{13}C NMR spectra because coupled spectra are too complicated to interpret due to large numbers of overlapping peak clusters. It is also the case that focusing each cluster into a single peak leads to taller peaks relative to noise (better signal to noise ratio).

27. **True**, in decoupled ^{13}C NMR each peak cluster is reduced to a singlet (a single peak).

28.

29.

30.

31.

32.

ChemActivtiy 17: Proton (^{1}H) NMR

Extend Your Understanding Questions (to do in or out of class)

17. See below.

18. The multiplicities of the peak clusters for H_a and H_b are note what would be expected based on the rules. H_a should be a triplet and H_b should be a multiplet (doublet of quartets).

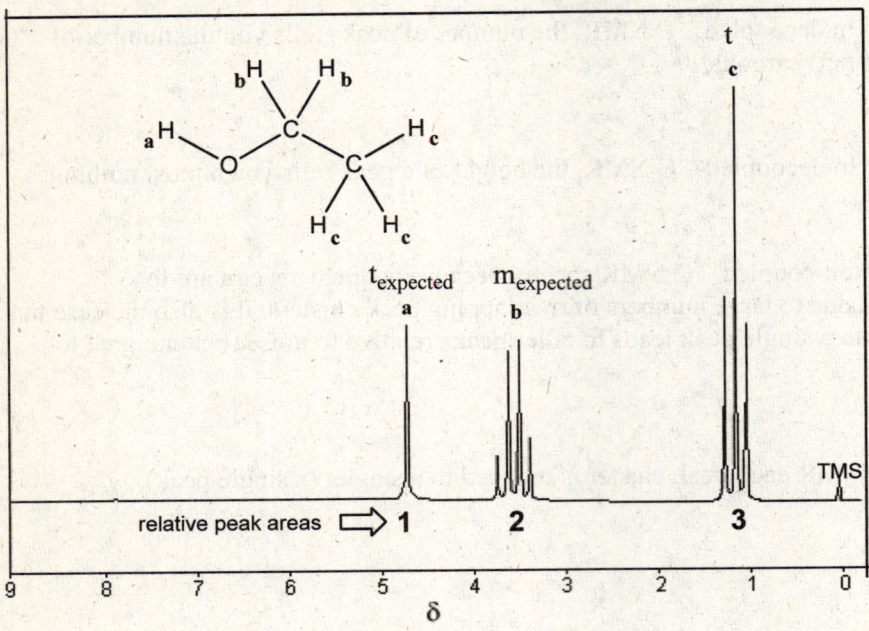

19. Memorization Task 17.5 explains the inconsistencies noted above because without coupling between H_a and H_b the spectrum is as shown above.

20. Any solvent with a hydrogen (such as $CHCl_3$, C_6H_6, CH_2O, etc.) is *not* useful as a proton NMR solvent because the solvent is usually present in much higher concentrations than the molecule being analyzed. This means that the signal(s) from the solvent hydrogens would dwarf the signals from the molecule being analyzed. The signals due to the analyte molecule would be tiny peaks hidden in the baseline noise.

21. Use of $CDCl_3$ (deuterated chloroform, or chloroform-d) consistently generates a singlet at 7.24 ppm. This peak can be used to calibrate the x axis of the spectrum, removing the need for addition of TMS to the sample.

22. Cross out the two molecules below that cannot serve as a proton NMR solvent.

Confirm Your Understanding Questions (to do at home)

23. From left to right on the spectrum, the peak clusters are assigned as follows: H_b at 3.6ppm, t (2H); H_a at 2.2ppm, s (1H); H_c at 1.6ppm, m (2H); H_d at 0.9ppm, t (3H)

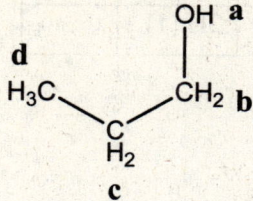

24. No

25. Yes, when D_2O is added to the sample, the only sample peak to disappear is the peak due to the alcohol hydrogen (OH). This is because only this H is exchanged with a D.

26.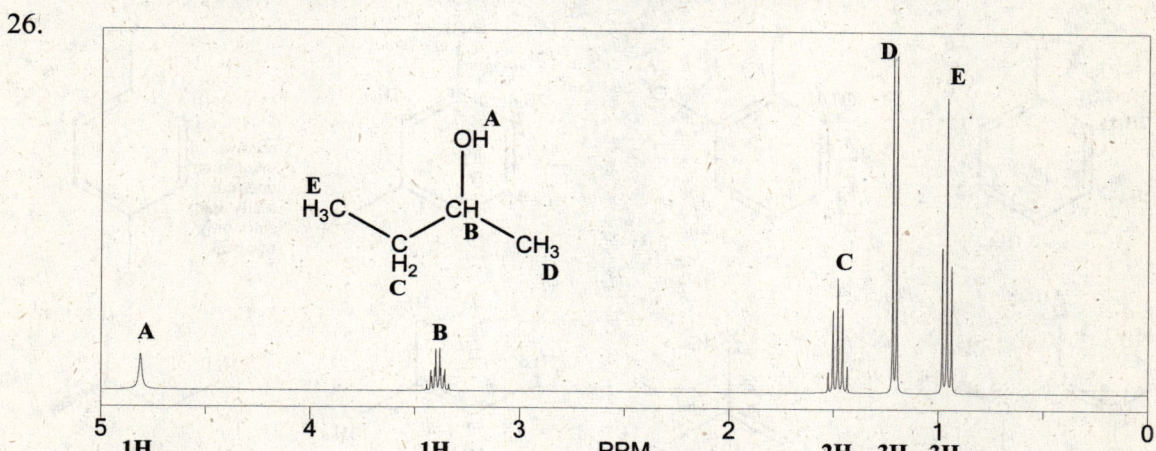

27.
   a. 3
   b. No
   c. Yes
   d. 5
   e. Yes
   f. When more than 4 peaks are present it is often hard to accurately count the number of peaks since the smallest peaks on the outside are often very small.

28.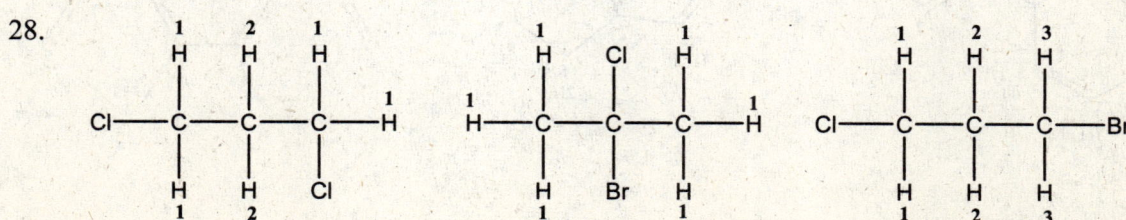

Continued on the next page. Relative integrations are shown in parentheses…

H type	Int	Mult
1	4H (2)	t
2	2H (1)	m (5)

H type	Int	Mult
1	6H (1)	s

H type	Int	Mult
1	2H (1)	t
2	2H (1)	m (5)
3	2H (1)	t

29.

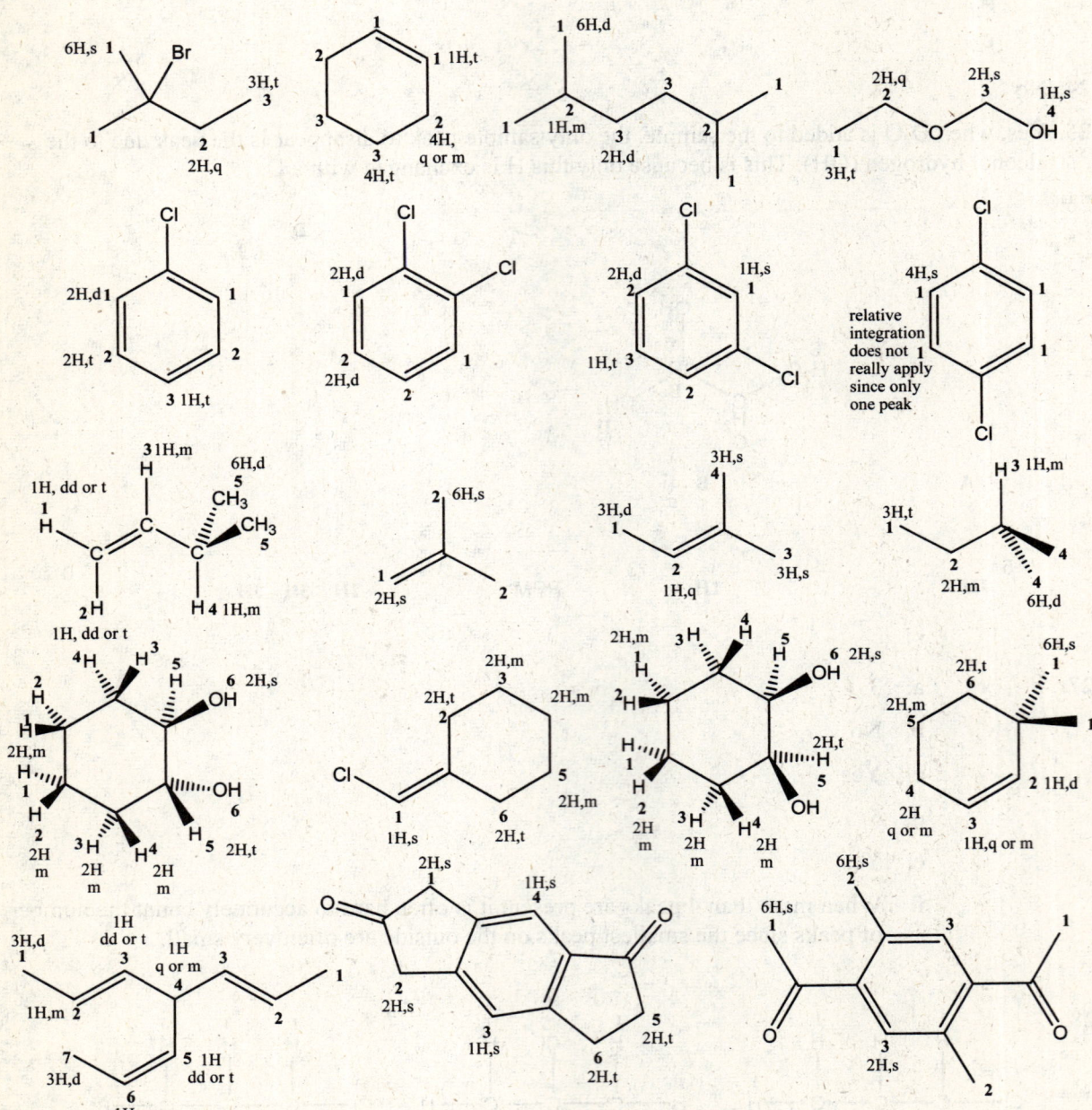

30. Two enantiomers will have the exact same NMR spectra. That is, you cannot distinguish two enantiomers using ordinary NMR techniques. (In practice, it is possible to add a chiral reagent, often an ammonium salt, to the mixture of enantiomers. This chiral salt interacts differently with each enantiomer, and generates two diastereomeric complexes with two different NMR spectra. *By analogy, imagine of a mixture of 50 left and 50 right mannequin hands. An analysis of the mixture using an instrument that cannot distinguish between enantiomers would simply "see" 100 hands. Now fit each hand with a RIGHT HANDED glove. Now this same instrument might see 50 hands in well-fitting gloves (the right hands), and 50 hands in ill-fitting gloves (the left hands).*

31. D_2O

32. It is an aromatic molecule (as indicated by the peaks between 6-9 ppm). Most likely it is a mono-substituted ring (based on the splitting pattern). The single substituent on the ring is an electron withdrawing group (based on the fact that the peaks are shifted to the left relative to benzene). It is a strong electron withdrawing group based on the fact that there is a big ppm difference between the three different peaks. (With a weak electron withdrawing or donating group, peak clusters due to H's on a benzene ring often overlap or are very close together.)

33. [structure: benzene ring with $-NO_2$ substituent]

34. Before trying to determine the structure you should first calculate that the molecule has one degree of unsaturation (1 ring or pi bond).

    [structure: butan-2-one]

35. a. see CTQ 17.
     b. same as for CTQ 17 but without the singlet

36. You can use MS data to determine the number of carbons and that there is a Br present. The peak near 10 ppm tells you that there must be an oxygen (aldehyde H around 10 ppm). This gives the molecular formula (C_3H_5OBr). Now use the NMR data to determine the structure...

    [structure: 2-bromopropanal]

37. The peak in the spectrum found at 4.5 ppm is outside the typical range for a hydrogen alpha to a carbonyl (2-3) *and* the typical range for a hydrogen bound to the same carbon as a halogen (3-4) because these effects tend to be somewhat additive. That is because this hydrogen is both alpha to a carbonyl and attached to the same carbon as a halogen it is expected to appear farther to the left than an H that is one or the other (but not twice as far to the left).

38. Use MS data to determine the number of carbons, that there is a Cl present, and finally the molecular formula ($C_{10}H_{11}Cl$). Then use the NMR data to determine the structure...

    [structure: 1-chloro-4-(2-methylprop-1-en-1-yl)benzene]

39.

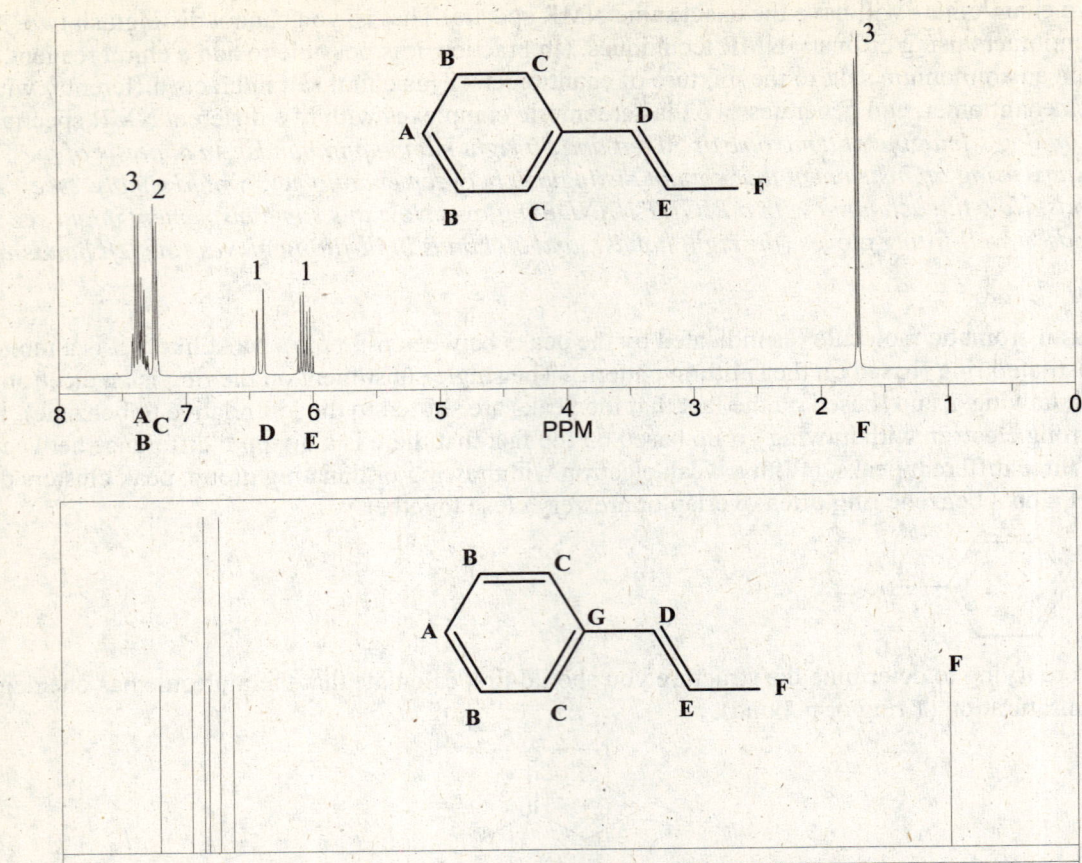

40. Proton NMR peaks generally are limited to between 0 and 11 ppm, while carbon NMR peaks usually fall between 0 and 220 ppm. Also, because coupled carbon NMR is essentially never used, the fact that the top spectrum shows splitting indicates that it is a proton NMR spectrum.

41. 7

42. The peak at 6.1 is expected to be a doublet of quartets because this H (marked "E" on the structure above) is split into a doublet by the H marked "D". Then each of these peaks is split into a quartet by the H's marked "F".

43.
    a. The circled H goes with the peak at 6.4 ppm (E).

    b. Splitting for peak E on the spectrum above is larger than on the spectrum below.

    c. The structure on the left goes with the spectrum above, and the structure on the right goes with the other spectrum based on the fact that the J value for the other spectrum is smaller, and a smaller J value is associated with the Z stereoisomer.

    d. The answer is the same as for the spectrum above, except that the Z stereoisomer should be used instead of the E stereoisomer that is shown above.

    e. See above.

Nomenclature Worksheet 1: Naming Alkanes and Cycloalkanes

22. There is only one way to place a methyl group on a three carbon chain so the designation that the methyl group is on C_2 is acceptable, but redundant.

23.

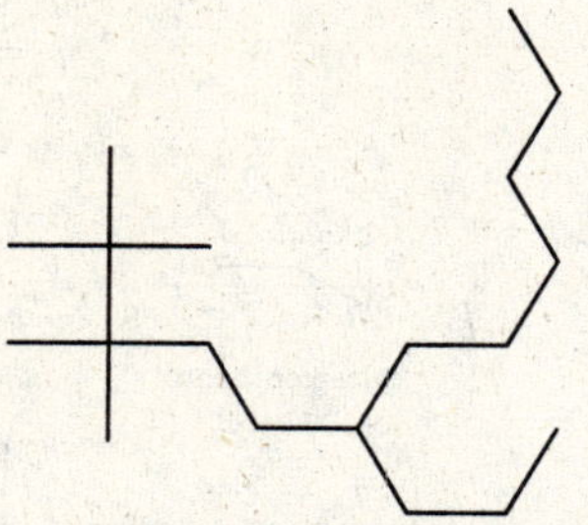

2,2,3,3-tetramethyl-6-propyldodecane

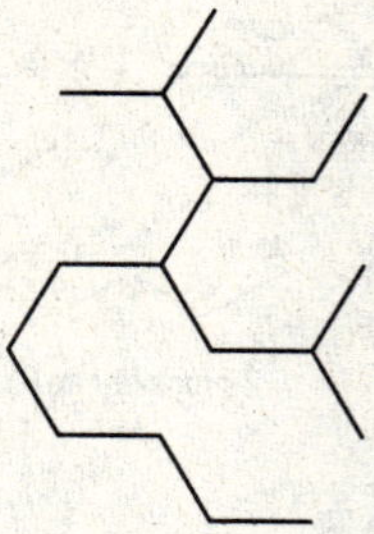

3-ethyl-4-isobutyl-2-methyldecane

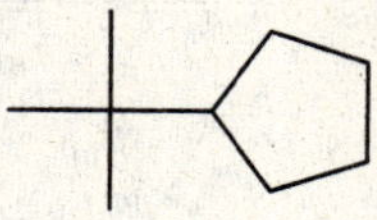

tert-butylcyclopentane

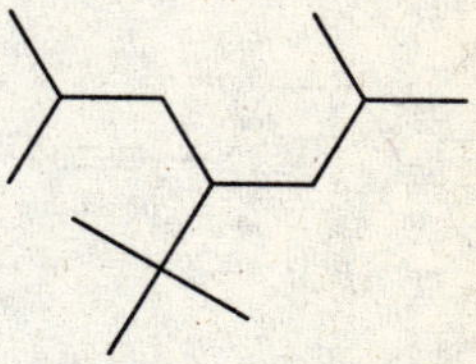

4-t-butyl-2,6-dimethylheptane

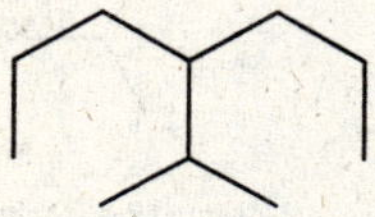

4-isopropylheptane

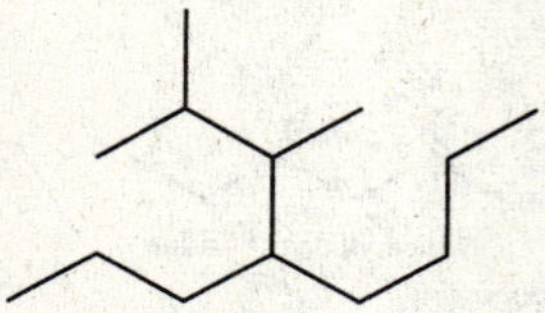

2,3-dimethyl-4-propyloctane

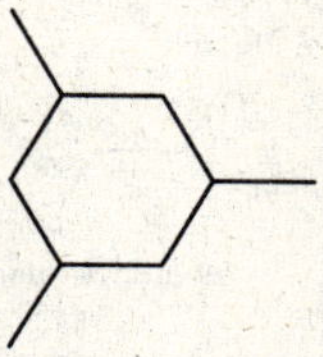

1,3,5-trimethylcyclohexane

24.

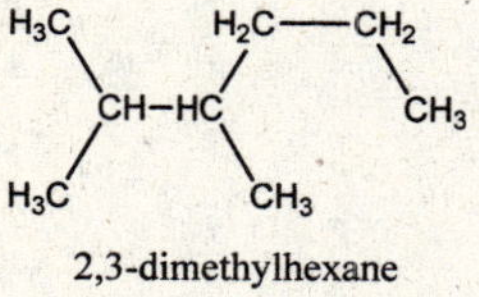

2,3-dimethylhexane

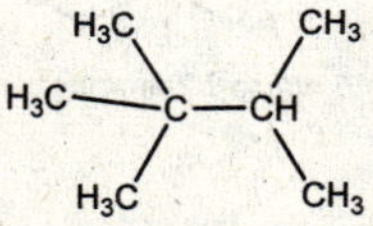

2,2,3-trimethylbutane

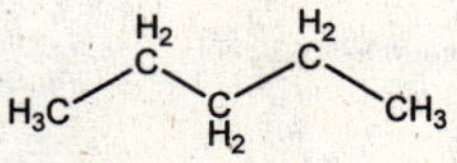

pentane

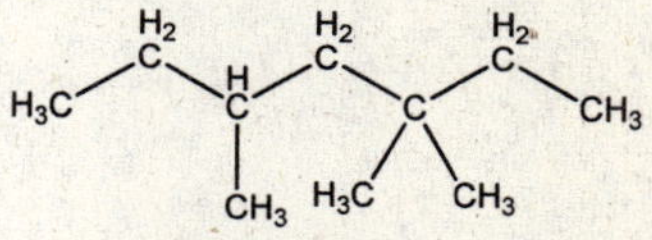

3,3,5-trimethylheptane

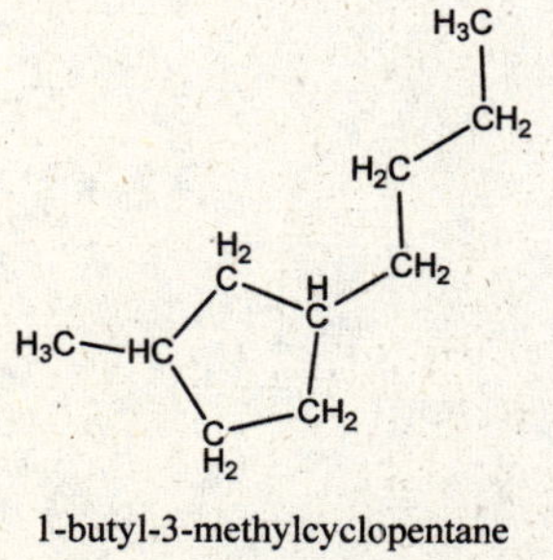

1-butyl-3-methylcyclopentane

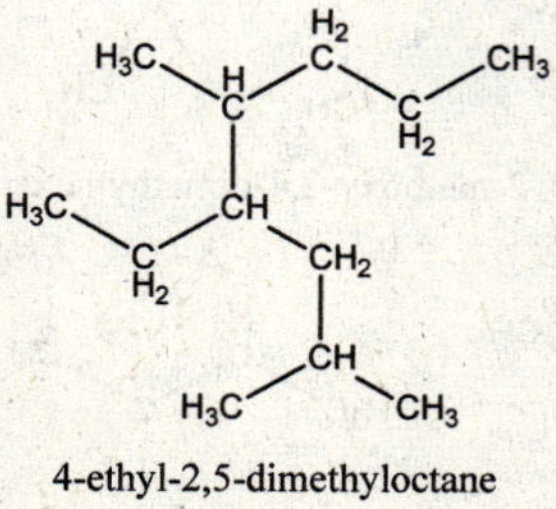

4-ethyl-2,5-dimethyloctane

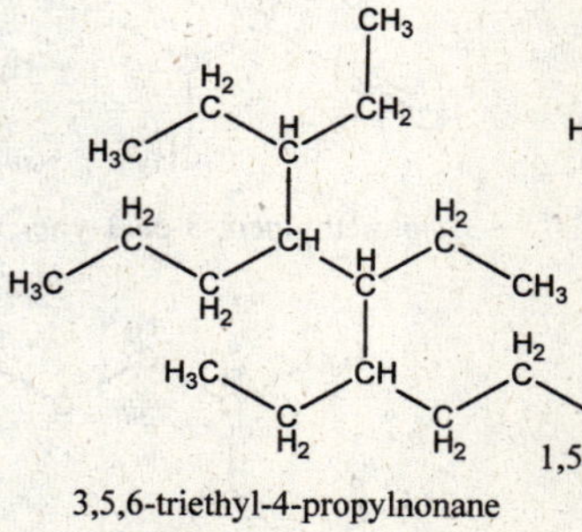

3,5,6-triethyl-4-propylnonane

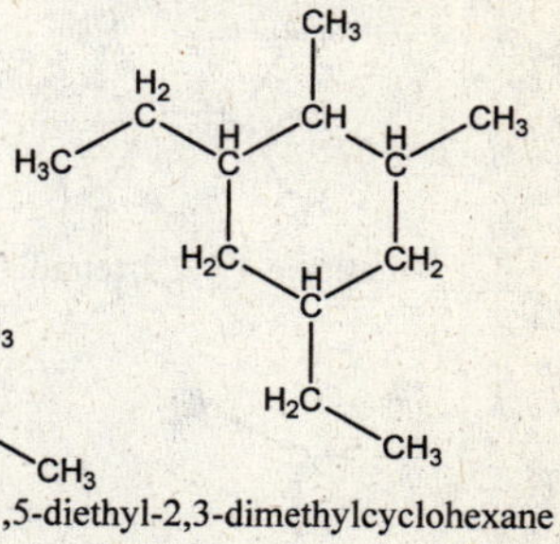

1,5-diethyl-2,3-dimethylcyclohexane

Nomenclature Worksheet 2: Intro to Naming Functional Groups

22.

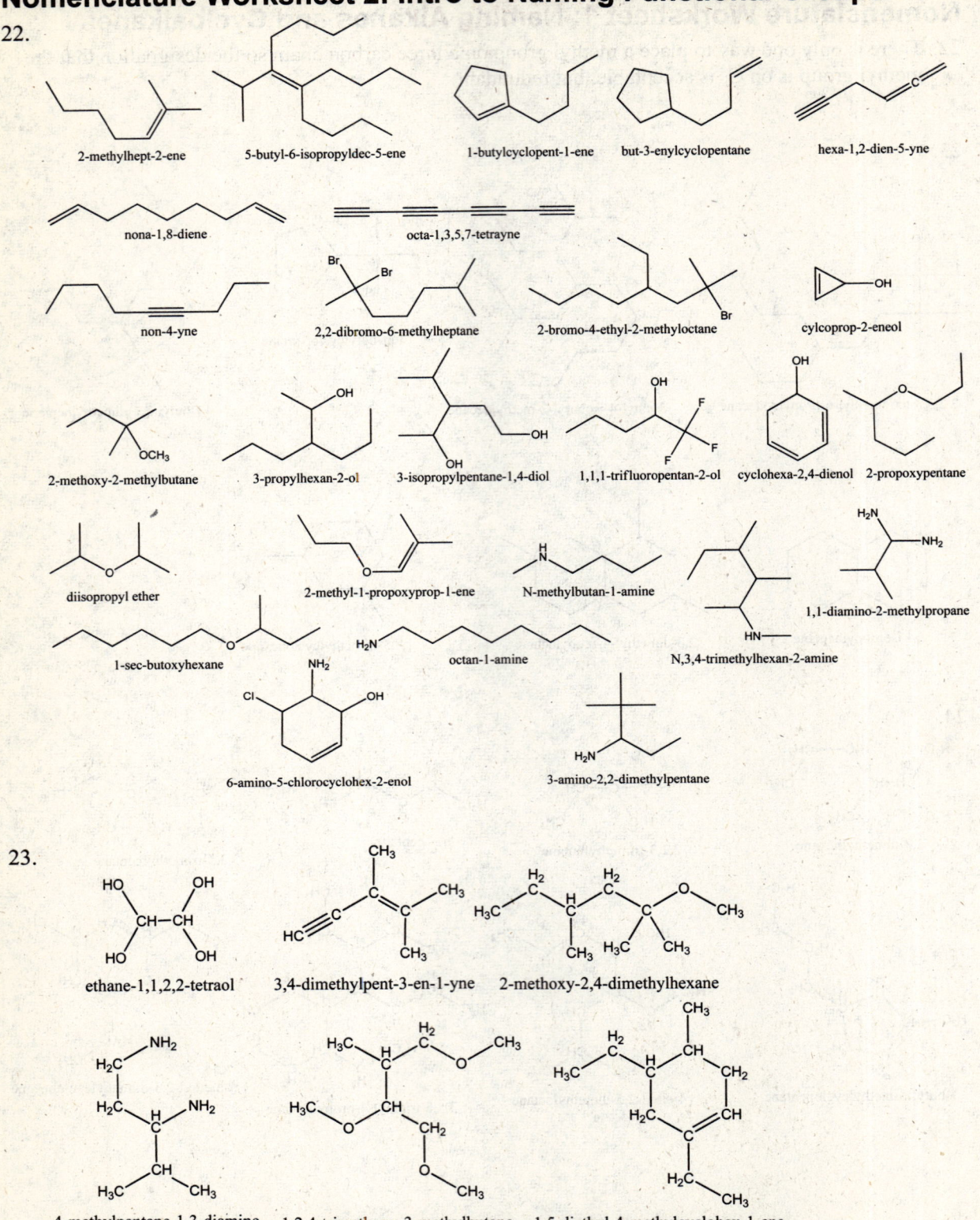

23.

ChemActivity 1: Aromaticity

Extend Your Understanding Questions (to do in or out of class)

10. *See below.*

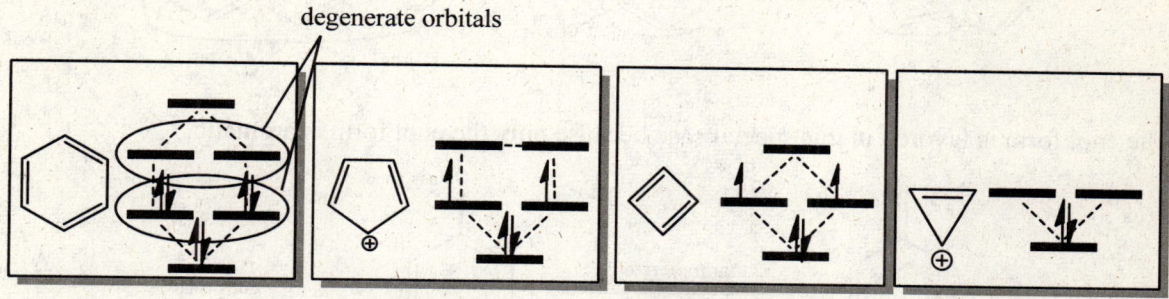

11. *See above.*

12. *See above.*

13. Cyclobutadiene is so unstable because it has two unpaired electrons (is anti-aromatic).

14. The key attributes of cyclobutadiene that make it anti-aromatic are that is has a cyclic, continuous, planar, pi system with 4n electrons.

15. Yes.

16. a. Cyclooctatetraene, if it were planar, would be anti-aromatic. By flexing out of planarity, the molecule adopts an less symmetrical molecular orbital diagram (with no degenerate orbitals). Since the molecule has no degenerate orbitals (as long as it has an even number of pi electrons) it will not be anti-aromatic.
    b. Cyclooctatetraene is neither aromatic nor anti-aromatic so it is non-aromatic.

17. Cyclobutadiene is one of the few molecules that is unable to "avoid" being anti-aromatic because it is one of the few molecules that is forced to be planar. (Larger rings can flex and become non-planar).

Confirm Your Understanding Questions (to do at home)

18. This molecule cannot be aromatic, so there is no reason for the N in this molecule to adopt an sp^2 hybridization state (since doing so would put 8 electrons in the pi system). Since there is no reason to adopt an alternate hybridization state, the N is expected to adopt the hybridization state that minimized electron pair repulsion. In this case, sp^3.

19. There are many correct answers to this question.

20. Only the one on the left is **aromatic**.

21.

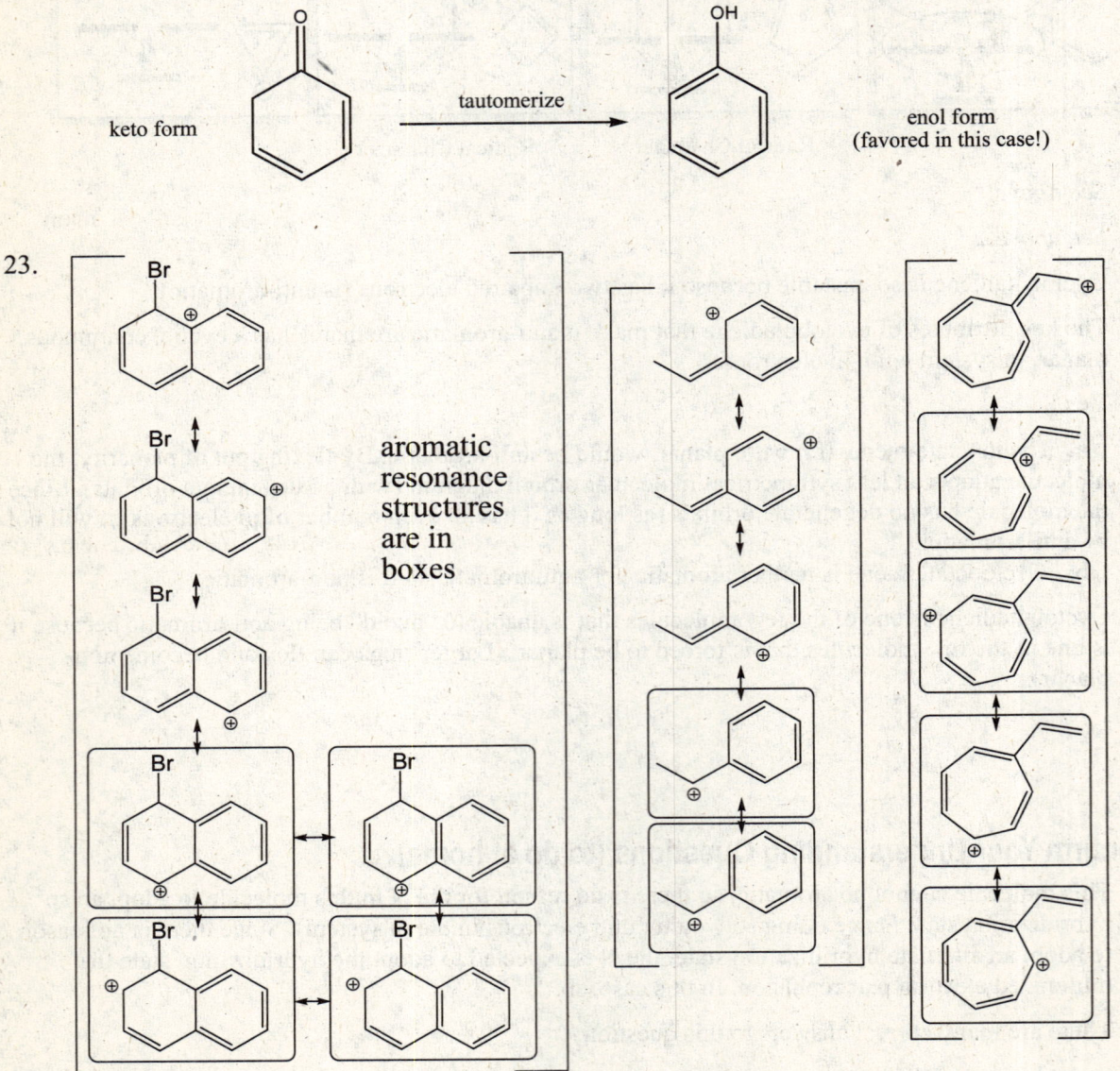

22. The enol form is favored in this special case because only the enol form is aromatic.

keto form → tautomerize → enol form (favored in this case!)

23. aromatic resonance structures are in boxes

24. From left to right: a) 4n pi electrons b) sp³ hybridized carbon interrupts all possible "racetracks"
c) sp³ hybridized carbon interrupts only possible "racetrack" d) 4n pi electrons e) 4n pi electrons
f) 4n pi electrons, though at least one resonance structure can be drawn that is aromatic.

25. This compound has a large amount of zwitterionic character because polarization of the C=O bond as shown both fits the natural polarization of a C=O bond AND generates a resonance structure that is aromatic. That is to say that by polarizing this bond the molecule takes on aromatic character and some of the low energy associated with aromaticity.

26. The lone pair on the left molecule resides in a p orbital (despite the fact that in a p orbital the lone pair resides closer (90° vs. 109.5°) to N-H and N-C sigma-bonding electrons) because placing this lone pair in a p orbital makes the molecule aromatic. In other words, it is worth the added energy to force the lone pair into closer proximity to other electrons because doing so gives the molecule the energy-lowering benefits of aromaticity.

27. All the molecules below are aromatic.

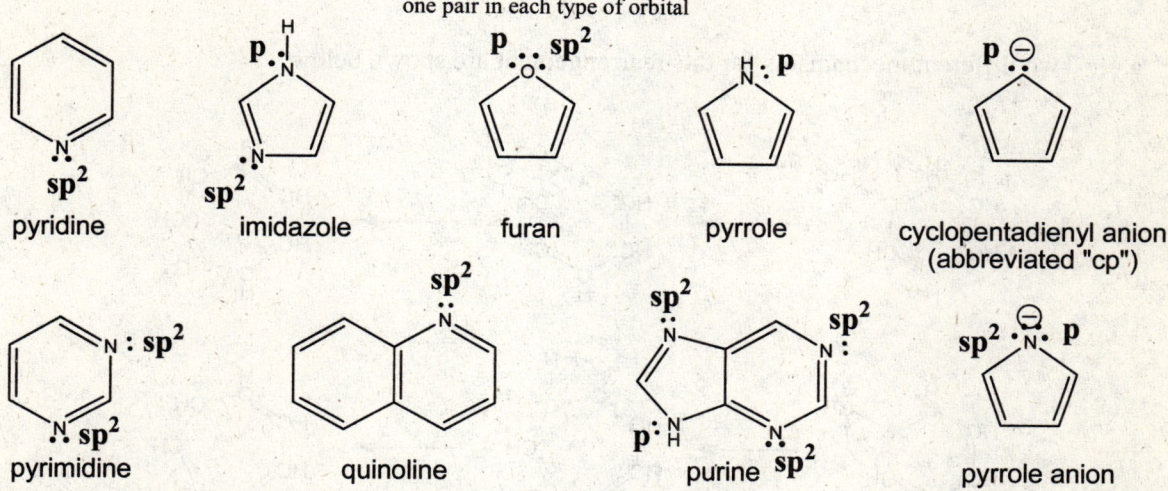

28. They are all aromatic and therefore expected to be extra low in energy and stable.

29. You can think of pK_a as the amount of energy required to remove an H from a molecule. The pK_a of cyclopentadiene is much lower than the pK_a of cyclopentane because the aromatic conjugate base of cyclopentadiene is lower in energy than the conjugate base of cyclopentane. In other words, you need to add less energy to cyclopentadiene to go UP to the energy of its conjugate base, because its conjugate base is closer in energy to the starting material energy.

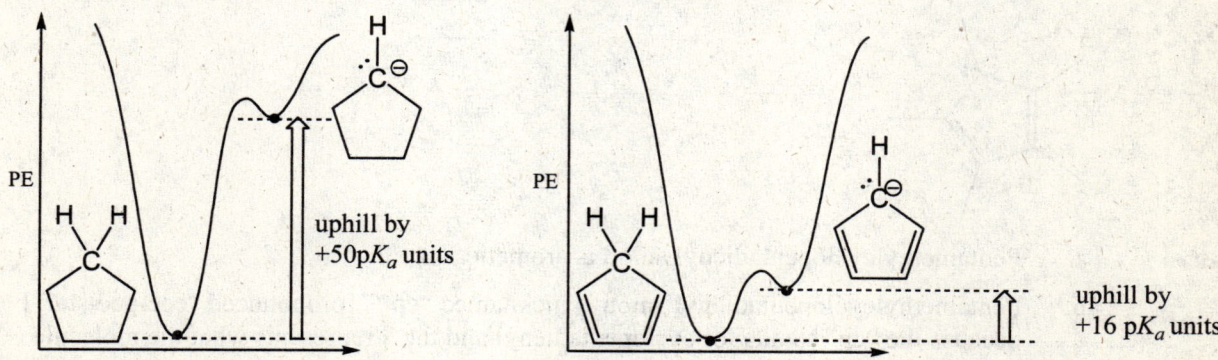

30. Carbocation A is formed first, but it quickly rearranges to Carbocation B. The driving force for this rearrangement is the fact that B is aromatic.

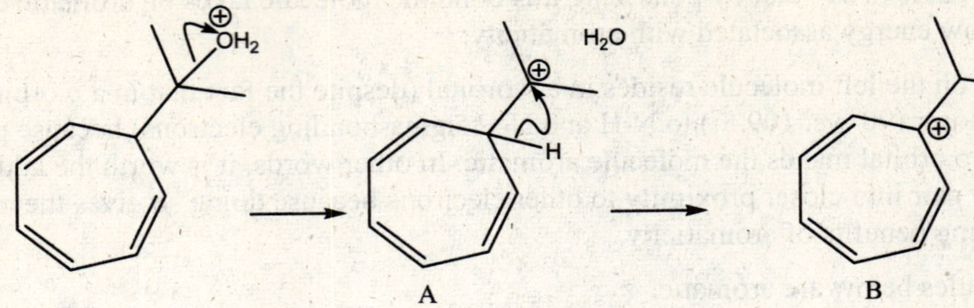

31. a. Two different mechanisms for this rearrangement are shown below.

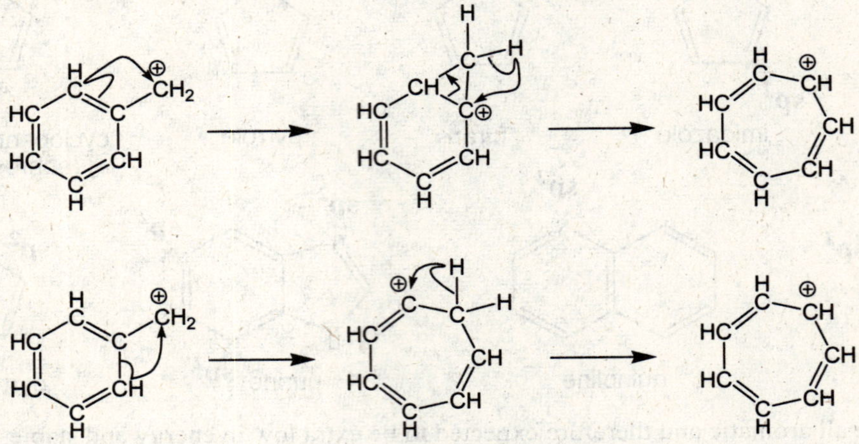

   b. The driving force for the rearrangement is NOT simply the fact that tropylium is aromatic (because the starting carbocation is aromatic as well). The driving force is the fact that tropylium is aromatic and *secondary*, whereas the starting carbocation is aromatic and primary.

32. 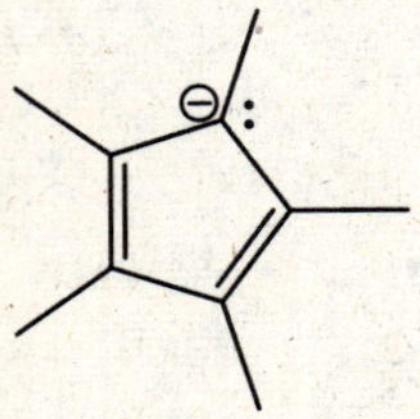
   a. Pentamethylcyclopentadienyl anion is aromatic.
   b. Pentamethylcyclopentadienyl anion is nicknamed "**cp***" [pronounced "cee-pee-star"] because the "cp" stands for cyclopentadienyl and the * represents what the molecule looks like, a five pointed star.

33.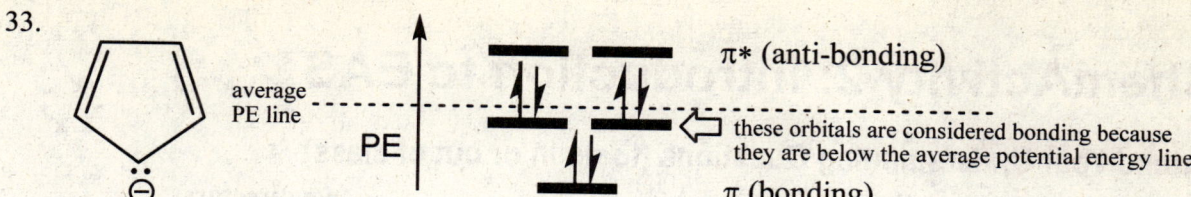

   a. Yes (when it had four electrons). Note that the diagram above now has six electrons.
   b. See above.
   c. The lone pair on this anion will reside in a *p* orbital because doing so gives it 4n+2 pi electrons, and makes it aromatic.
   d. Yes.

34.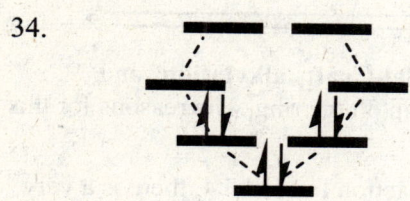

35. According to MO theory, a cyclic conjugated π system with 4n electrons (4, 8, 12, etc.) will be very unstable because it will have two unpaired electrons. The fact that all such systems will have two unpaired electrons is a consequence of the common shape of MO diagrams for these types of cyclic molecules. They all have one lowest PE molecular orbital followed by one or more pairs of degenerate (equal energy) molecular orbitals. Since molecular orbitals of the same energy must both be half filled before either is filled, this leads to two unpaired electrons when the total number of pi electrons is 4n. These unpaired electrons give the molecule radical character, which makes it very reactive (like a radical).

36. a. The conformation on the left is very unfavorable because the bond angles are very far away from the optimal 120 degrees expected for sp^2 hybridized atoms.
    b. This molecule is non-aromatic because it cannot be planar (due to the fact the two H's shown would have to occupy the same space in order for the molecule to be planar).

37. It is impossible for this molecule to have a pi system with 10 electrons. Though the three pi bonds are conjugated, only one of the lone pairs on oxygen can participate in this pi system. The other lone pair will reside in an orbital that is perpendicular to the p orbitals of the conjugated pi system. This molecule will adopt a non-planar conformation and will be nonaromatic.

ChemActivity 2: Introduction to EAS

Extend Your Understanding Questions (to do in or out of class)

10.

m directors — Electron-Withdrawing Groups (EWG) — very strong EWG's → weak EWG | weak EDG ← very strong EDG's — Electron-Donating Groups (EDG) — *o/p* directors

11. Be sure to note that Synthetic Transformations 2.5 and 2.6 "**Friedel-Crafts alkylation**" and "**acylation**," do not work with a strong electron withdrawing group on the ring. The reasons for this are not covered in this activity.

12. When R = isopropyl (as compared to when R = methyl) for the reaction in Model 4, there is a very strong preference for the electrophile to substitute at the *para* position because the *ortho* position is partially blocked by the adjacent isopropyl group, significantly slowing substitution at the *ortho* position.

13. Without steric effects, one would expect the *ortho* product of the reaction to be exactly twice as abundant as the *para* product because there are two different positions on the ring that lead to the *ortho* product, but only one position on the ring where substitution leads to the *para* product.

14.

15. electrophile is made from a reaction of SO_3 and sulfuric acid

16. [mechanism showing three resonance structures of arenium ion intermediate with H and D substituents]

17. [mechanism showing sulfuric acid protonating nitric acid to generate nitronium ion NO₂⁺]

18. TNT

19. [mechanism of Friedel-Crafts alkylation of benzene with ethyl cation, showing arenium intermediate, deprotonation by AlCl₄⁻, and formation of ethylbenzene]

20. [mechanism of Friedel-Crafts acylation: AlBr₃ activating acyl bromide of cyclopentanecarbonyl bromide to form acylium cation (two resonance structures), then benzene attack, arenium intermediate, deprotonation by AlBr₄⁻ to give phenyl cyclopentyl ketone plus AlBr₃ and HBr]

21. a. Br₂ reacts with cyclohexene but not with benzene because cyclohexene is a much better nucleophile than benzene. The explanation for this is that the pi bond on cyclohexene is isolated (not even resonance stabilized). In comparison, each pi bond on benzene is highly resonance stabilized (so much so that the special term aromatic is assigned to this system). The result is that benzene is relatively unreactive, and requires a very potent electrophile to start the reaction. A Lewis acid catalyst such as $FeCl_3$ makes Cl_2 a more potent electrophile.

Continued on next page.

b.

22. One way to explain that the one with the alkyl group is faster is to cite that alkyl groups are electron donating, which makes the alkyl benzene ring more electron rich than the unsubstituted benzene ring. Since the ring functions as a nucleophile in EAS, making the ring more electron rich makes it a better nucleophile. A rigorous explanation of why a reaction is faster must talk about the transition state specifically, so you could add to the above explanation that the better nucleophile makes a lower energy transition state in the slow step (1st step in this case) of the reaction. Another way to explain this is to use the Hammond postulate and state that the pathway with the lower energy intermediate will have the lower energy transition state. The transition state for the alkyl benzene has tertiary carbocation character (one of the three resonance structures is tertiary, while all three r.s. of the intermediate with plain benzene are secondary carbocations). This means the intermediate on the alkyl benzene pathway is lower in energy, so this pathway is expected to be faster.

23.
   a. [benzaldehyde structure]
   b. [anisole structure]

24.
   a. The pi system of the ring acts as [**a nucleophile**] in an EAS reaction.
   b. The [**more**] electron-rich the pi system of the aromatic ring, the faster the rate of EAS.

25. Benzene will undergo EAS reaction faster than nitrobenzene because of the answer to the previous question. Benzene is more electron rich, and so is a better nucleophile in EAS.

26. Each nitro group is a powerful electron withdrawing group so having two of these directly attached to the ring deactivates the ring as a nucleophile in EAS.

27. Toluene "hogs" all the D-Cl while the nitrobenzene does not get any because toluene is much more electron rich than nitrobenzene. This makes it a better nucleophile and allows it to out-compete nitrobenzene for the available electrophile. Basically, the reaction with toluene is so fast compared to the reaction with nitrobenzene that the toluene reaction uses up all the electrophile before much nitrobenzene-derived product can be formed.

ChemActivity 3: EAS Resonance Effects

Extend Your Understanding Questions (to do in or out of class)

10. The second-order resonance structures of phenol demonstrate that the *ortho* and *para* positions are more electron rich than the *meta* positions. Since the ring carbon acts as a nucleophile (using pi electrons) in an EAS reaction, these more electron rich positions are more likely to make a favorable transition state leading to a bond with the electrophile. For this reason, the EAS reaction in this question yields mostly the *ortho* and *para* products.

first order resonance structure   *second order*   *second order*   *second order*

(Note that a more complete explanation for why the ortho and para products are preferred must address the relative energies of the carbocation intermediates. Questions that focus on this issue can be found later in this homework.)

11. Any benzene ring with an electron donating group is expected to preferentially direct an electrophile to the *ortho* and *para* positions. See CA2, Model 2.

12. Because resonance electron-withdrawing and donating effects are usually strong …

    a. A resonance electron-withdrawing group is a strong **meta director**.

    b. A resonance electron-donating group is a strong **ortho/para director**.

13. When Z = OH, a higher yield of the di-chloro and tri-chloro products will be observed since OH is a stronger electron donating group than CH_3, and therefore a stronger activator of EAS.

14. The result of this reaction is a mix of multi-substituted alkyl benzene products and a large amount of unreacted starting material (benzene) because an alkyl group is an EAS activator, so each time an alkylbenzne product is produced, this product becomes a more attractive target for further EAS than any given benzene molecule.

15. In contrast to the previous question, the result of this reaction is the mono-substituted product because an acyl group is an EAS deactivator. That is, the product is far *less* likely to react with an electrophile than the starting material (benzene).

16. Yes.

17.

18.

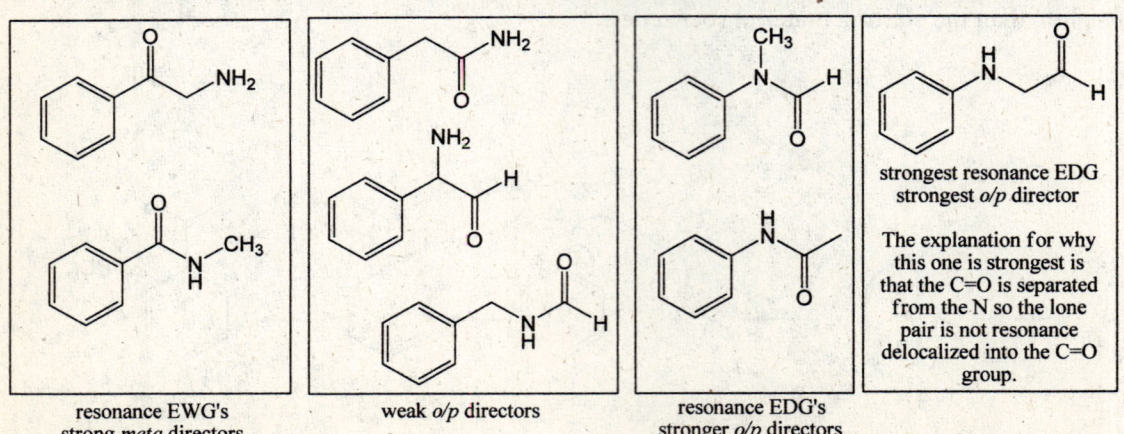

Confirm Your Understanding Questions (to do at home)

19. The basic rule is that Zn(Hg)/HCl will reduce any C=O group without reducing C=C bonds, and H$_2$/Pd will reduce any ordinary (non-aromatic) carbon-carbon double bond and any <u>benzylic</u> C=O.

20.

21. The enol of each structure below is not included because in each case the keto form is favored.

resonance EWG's
strong *meta* directors

weak *o/p* directors

resonance EDG's
stronger *o/p* directors

strongest resonance EDG
strongest *o/p* director

The explanation for why this one is strongest is that the C=O is separated from the N so the lone pair is not resonance delocalized into the C=O group.

22.

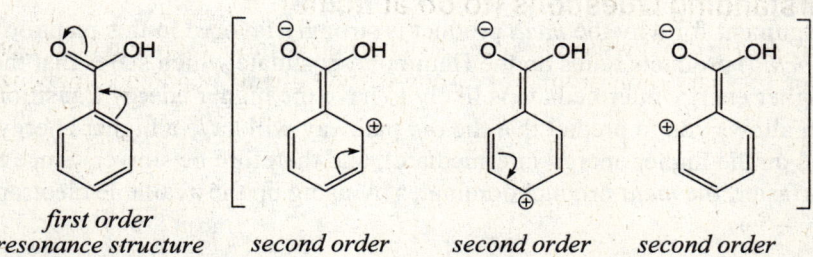

a. The second-order resonance contributors are less important than first-order resonance structures, and contribute only a small amount to our overall understanding of phenol based on the fact that the second order resonance structures have a much less favorable arrangement of charge. The criterion for second order versus first order is that second order resonance structures have a larger total number of formal charges.

b. Second-order resonance structures demonstrate that the *ortho* and *para* positions are more electron rich than the *meta* positions. Since the ring carbon acts as a nucleophile (using pi electrons) in an EAS reaction, these more electron rich positions are more likely to make a favorable transition state leading to a bond with the electrophile.

23.

a. Resonance-withdrawing group

b. The second-order r.s. demonstrate these positions are electron deficient. Since the ring acts as a nucleophile (using pi electrons) in an EAS reaction, these less electron rich positions are less likely to make a favorable transition state leading to a bond with the electrophile.

24.

a. With nitrobenzene, the electrophile is directed to the *meta* position, whereas with toluene or aminobenzene (aniline) the electrophile is directed to the *ortho* or *para* positions.

b.

c.

d.

e. The intermediate on the reaction pathway to the *meta* product is lowest in potential energy because it has no highly unfavorable resonance structures with two positive charges next to one another.

Draw an energy diagram showing all three pathways (*ortho, meta* and *para*).

f.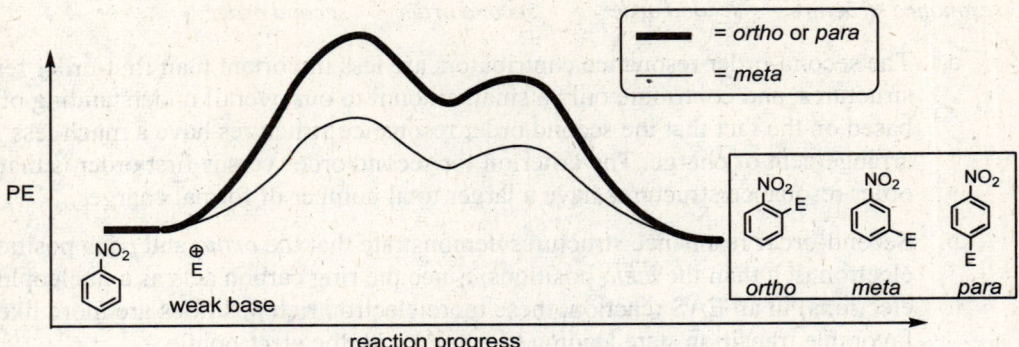

g. The best argument for why the *meta* product is strongly favored in this reaction over the *para* (and *ortho*) products relies on the Hammond postulate which states that the pathway with the higher energy intermediate is likely to have the higher energy transition state. This assumption allows you to predict that the *o/p* pathway will have a higher energy transition state (based on the higher energy intermediate), and therefore be slower. Since the *meta* pathway is faster, the *meta* product dominates by using up the available electrophile first.

25. a. [resonance structures of meta intermediate]

Note: The *ortho* intermediate, shown below, is similar to the *para* intermediate.

b. [resonance structures of ortho intermediate]

c. You can argue that the intermediate on the pathway to the *meta* product is higher in potential energy than the intermediate on the pathway to the *ortho* or *para* products based on the fact that the *ortho* or *para* intermediates are lower in energy thanks to more resonance stabilization (more resonance structures).

d.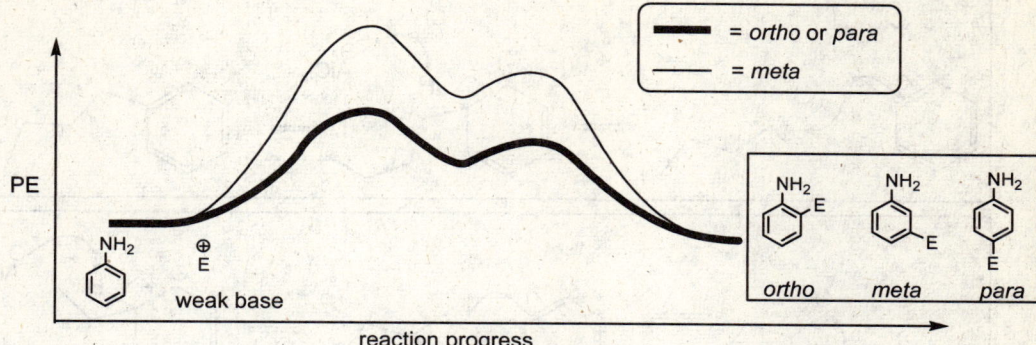

26.

a. One explanation is that resonance effects are generally stronger than inductive effects. Lets compare both reactions to an EAS reaction with benzene. A resonance argument explains why the intermediate on the o/p pathway for aniline is lower in energy than that of benzene, whereas the tertiary carbocation character of the EAS intermediate on the o/p pathway for toluene is used to explain why that pathway is faster than the benzene pathway. (Note that the fact that tertiary carbocations are lower in energy than other carbocation is considered an inductive effect. That is, alkyl groups are inductively electron donating.)

b. Based on the Hammond postulate we expect the reaction with the lower energy intermediate to be faster. In this case, aniline.

27. One explanation for why an ammonium ion group ($^+NR_3$) is a *meta* director is that the circled resonance structure below is very unfavorable due to the proximity of the two positive formal charges. This indicates that the intermediate on the *ortho* (or *para*) pathway is unfavorable and high in energy, and so the nearby transition state is also likely to be high in energy.

28.

29. a) (1-bromo-1-phenylpropane: PhCH₂CH(Br)CH₃) b) (2-bromo-2,3-dimethylbutane)

30. In pathway a there is going to be some rearrangement leading to a product with a *tert*-butyl group instead of the desired alkyl side-chain.

31.

32. Only nitrobenzene will be formed. This is because benzene undergoes EAS much faster than nitrobenzene so if there is one equivalent of electrophile, all the benzene will use it up before a noticeable about of dinitrobenzene could form.

33. In contrast to the previous question, ethylbenzene undergoes EAS faster than benzene. This means that as soon as a significant concentration of ethylbenzene is present in the solution, it will begin to out-compete benzene for the remaining molecules of electrophile. The result will be a large amount of diethyl benzene, and even some triethylbenzene.

34.

35. The reason why using a huge excess of benzene will reduce the likelihood that polyalkylation products will form is that at every moment during the reaction (until electrophile is used up) each electrophile is likely to be surrounded by a large number of benzene molecules. As long as the excess of benzene outweighs the speed difference between the competing reactions, the reaction with benzene can be the major pathway.

36. The basic rule is that Zn(Hg)/HCl will reduce any C=O group without reducing C=C bonds, and H_2/Pd will reduce any ordinary (non-aromatic) carbon-carbon double bond and any benzylic C=O.

Note that Zn(Hg)/HCl will reduce a benzylic C=O more easily than a non-benzylic C=O so careful control of temperature and time could allow selective reduction of the benzylic C=O.

H_2/Pd can reduce almost any pi bond given a high enough temperature and pressure. Even the pi bonds of a benzene ring can be reduced at very high temperature and pressure. However, at normal temperature and pressure H_2/Pd will reduce ordinary C=C pi bonds and benzylic C=O bonds to CH_2, but have no effect on a non-benzylic C=O bond. (Note that at moderate temperatures and pressures H_2/Pd can be used to reduce a non-benzylic C=O bond to a C—OH bond; however, $LiAlH_4$ or $NaBH_4$ are normally used to reduce a carbonyl to an alcohol.)

37.

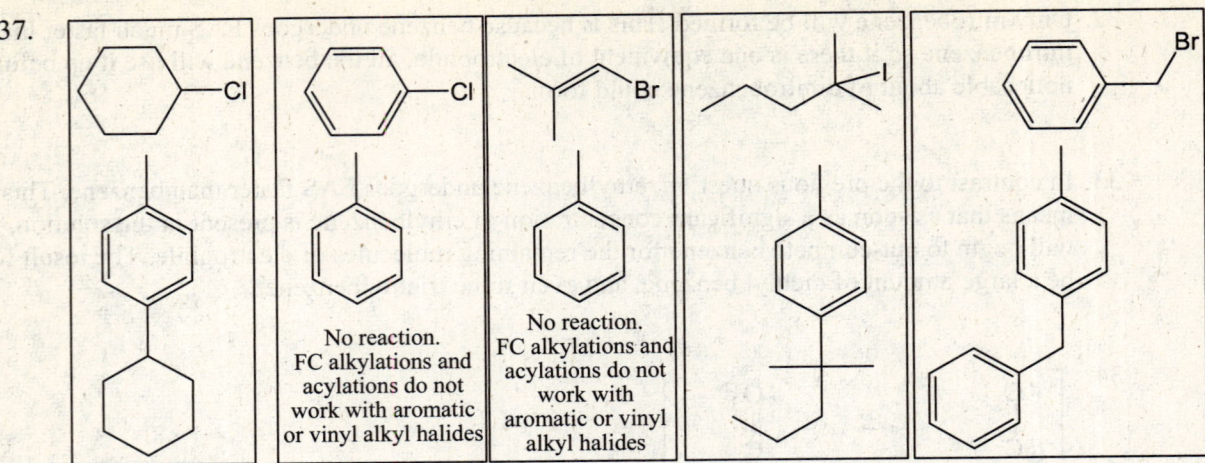

38. Benzene is far more likely to react with the electrophile then either of the substituted benzene rings (giving nitrobenzne). This is because the substitutent in this case is a strong electron withdrawing group which deactivates the ring toward further substitution. The significance of this question lies in the fact that if benzene is treated with sulfuric and nitric acids the result is mono-substitution of the ring (dinitrobenzene is not observed). This is in contrast to the reaction in the next question.

39. In stark contrast to the previous question, in this mixture, the most substituted ring (diethylbenzene) is most likely to react with the electrophile (giving triethylbenzne). The reason is that an alkyl group is an activator in EAS, so the more substituted ring is more highly activated toward further reaction. The significance of this question lies in the fact that treatment of benzene with $EtCl/AlCl_3$ is NOT a clean reaction, and leads to multiple products. As soon as one molecule of ethylbenzene is produced, this product is more likely than a given molecule of benzene to undergo a second alkylation. Then, as soon as dielthyl benzene is found in the reaction mixture it will begin to out-compete benzene for an electrophile.

ChemActivity 4: EAS Competing Effects

Extend Your Understanding Questions (to do in or out of class)

8. This reaction yields the product shown, and not a product with the Br *ortho* to the *tert*-butyl group because the *tert*-butyl group blocks the Br from substituting in the *ortho* positions.

9. If you replace the reagents $Br_2/FeBr_3$ in Rxn A Model 2a and Rxn D in Model 2b with $CH_3Br/AlBr_3$ the result is no reaction since Friedel-Crafts reactions do not work on strongly deactivated rings.

10. Yes.

11. This reaction gives no product (or almost no product) since two strong deactivators essentially stops an EAS reaction.

12. Yes.

13. In the column labeled **Product Regiochemistry** (from top to bottom) the following labels apply: o/p; o/p; o/p; —; o/p; m; m; m; m; m.

14. Ammonium is a *meta* director but <u>not</u> a strong resonance electron withdrawing group.

15. Halogens are *ortho/para* directors that are not net activators.

Confirm Your Understanding Questions (to do at home)

16.
    a. Product **c** is not formed in Rxn I because, according to the rule in Mem. Task 19.15, any activator (even the weakest activator, an alkyl group) beats any deactivator (even one of the strongest deactivators such as a nitro group). This example is perhaps the toughest test of 19.15, and in practice this reaction gives a messy mixture of products, but to keep things manageable assume that 19.15 holds true in all cases.

    b. Product **d** is not formed in Rxn I because of steric factors.

    c. Product **f** is not formed in Rxn II because the directing effects of an activator outweigh the directing effects of a deactivator.

17.

    a. The two products above are formed for the same reasons as described in the previous question, parts a and b.

    b. The first product shown will dominate because of steric factors (CO_2H, especially with a molecule of $FeCl_3$ attached to it, is larger than CH_3).

18.
    a. T: A strong activator overpowers the directing effects of a weak activator.

    b. T: A weak activator overpowers the directing effects of a deactivator.

    c. T: All activators are *o/p* directors.

    d. F: All deactivators are *m* directors. Halogens are the sole exception to this rule.

19. (2) is most important; but (1) is also helps explain this.

20.
    a. The large size of a *tert*-butyl group blocks access by the electrophile to the ortho positions.

    b. Reaction III is closest to the expected 1:2 ratio because a methyl group has the smallest steric effects, and it is steric effects that cause the product ratio to stray from the 2:1 ratio expected based on the statistics of possible substitution positions.

ChemActivity 5: EAS Synthesis Workshop

Extend Your Understanding Questions (to do in or out of class)

12. *See below, left.*

13. *See structure in brackets, below right.*

14. Unlike aniline, an acylated aniline can undergo a single controlled EAS reaction because the acyl group reduces the electron donation by the nitrogen. This reduces the likelihood of multiple substitutions, prevents the nitrogen itself from acting as a nucleophilie and making a bond to the eletrophile, and (the acyl group) helps block the ortho position, making *para* substitution more likely than *ortho*.

15.

16.

17. The sulfonic acid group in the synthesis in the previous question is said to be used as a **protecting group** because it protects the preferred *para* position, temporarily, while the nitro group is forced to go to the more sterically hindered *ortho* position.

18.

| fast EAS | moderate NAS | neither | fast NAS | moderate EAS | neither |

19. Both start and end with an aromatic molecule, and the rate of each reaction is dramatically affected by the presence and placement of EWG's or EDG's on the ring. The key differences are that in EAS the ring is nucleophilic and activated by *ortho* and *para* EDG's while in NAS the ring is electrophilic and activated by *ortho* and *para* EWG's. Note also that in EAS an electrophile substitutes for an aryl hydrogen, while in NAS a nucleophile substitutes for an aryl halogen.

Confirm Your Understanding Questions (to do at home)

20. *tert*-Butyl benzene does not react with KMnO$_4$ because a benzylic hydrogen (or OH or C=O or other functional group) is necessary for this oxidation to take place.

21. Zn(Hg)/HCl will reduce any C=O group without reducing ordinary C=C bonds. (Reduction of a benzylic C=O is easier than a non-benzylic C=O so careful control of temperature and time could allow selective reduction of a benzylic C=O.)

H$_2$/Pd can reduce almost any pi bond given a high enough temperature and pressure. Even the pi bonds of a benzene ring can be reduced at very high temperature and pressure. However, at normal temperature and pressure H$_2$/Pd will reduce ordinary C=C pi bonds and benzylic C=O bonds to CH$_2$, but have no effect on a non-benzylic C=O bond. (Note that at moderate temperatures and pressures H$_2$/Pd can be used to reduce a non-benzylic C=O bond to a C—OH bond; however, LiAlH$_4$ or NaBH$_4$ are normally used to reduce a carbonyl to an alcohol.)

22.

23. Each of the following could be made in a number of different ways. The synthesis shown is not necessarily the best method. If you think you have a better method, share your synthesis with your group, TA, or instructor!

24. Make up a synthesis problem that focuses on synthetic transformations involving aromatic molecules. If it is a good synthesis problem, email it to your instructor (write a description of it in words), and there is a chance it will show up on the next quiz.

25. A leaving group is required for a NAS reaction. Neither of these rings has a leaving group.

ChemActivity 6: Organometallic Reagents

Extend Your Understanding Questions (to do in or out of class)

9.

10. See below.
11. See below.

Confirm Your Understanding Questions (to do at home)

12. a) R-CH₂-OH b) R-CH(OH)-R c) R—CN d) R-CH=CH-R

13. (CH₃)₃CH → NBS → Li → D₂O → (CH₃)₃C-D

14. PhBr → Mg → acetone → H⁺/heat → α-methylstyrene

15. a. [Mechanism diagram showing HO-CH₂-CH₂-Br + Mg metal → carbanion with ⁺MgBr, with dashed annotations: "bond to H⁺ releases −35 pKa units" and "requires +16 pKa units to break", noting "starting alkyl halide is in excess in the reaction mixture", leading to side-reaction products HO-CH₂-CH₃ and ⁻O-CH₂-CH₂-Br]

b. Preparation of lithium and Grignard reagents **cannot** be performed with an alkyl halide containing an H with a pKa < 35 because acid base reactions such as the one above interfere with preparation and use of the desired organometallic reagent.

16. In nucleophilic addition, a new reagent is added to the electrophilic carbon, but no group leaves. (A bond is broken, but it is only the pi portion of a double bond, so there is no leaving group.) In nucleophilic substitution the nucleophile displaces a leaving group. This is possible because it takes place at an sp³ hybridized carbon and a single bond is broken, liberating the leaving group.

Volume 2, ChemActivity 6: Organometallic Reagents

17. All except the last one, chlorocyclohexane, are unacceptable alkyl halides for making a Grignard or lithium reagent.

18.

19.

20.

ChemActivity 7: Addition-Elimination

Extend Your Understanding Questions (to do in or out of class)

10.

11.

12. a. Push the reaction to the right by removing water.
    b. Push the reaction to the left by adding water.

13. According to Model 3, an aldehyde/ketone is more susceptible to reaction with a nucleophile (e.g., RNH_2, H_3C-Li, $RC\equiv CNa$, $KC\equiv N$ than a cyclic acetal.

14. This reaction will NOT give a significant yield of the target shown because the nucleophile is much more likely to react with the aldehyde than with the ketone, resulting in formation of this product.

15. Yes. This data confirms that the first equivalent of nucleophile will react with the more reactive electrophile, the aldehyde.

16. An aldehyde carbonyl is more reactive than a comparable ketone carbonyl.

17. Because the aldehyde is more reactive than the ketone toward nucleophiles, the cyclic acetal is much more likely to form at the aldehyde carbon since it forms via a nucleophilic addition-elimination reaction.

18. Design a way to accomplish the following synthesis (from the previous page). *Hint*: The structure in the box above is formed during this synthesis.

    starting material → one equiv. $HOCH_2CH_2OH$, H^+ → CH_3Li → $NaOH_{aq}$ → neutralize → target

19. The molecule ethan-1,2-diol is considered a protecting group in the synthesis above because it temporarily protects the more reactive aldehyde carbon, forcing the methyl nucleophile to react with the ketone.

Confirm Your Understanding Questions (to do at home)

The answers on the next several pages have been used to showcase several valid ways to show a proton transfer in a mechanism. Any of these methods could be used to show any of these types of mechanisms.

20.

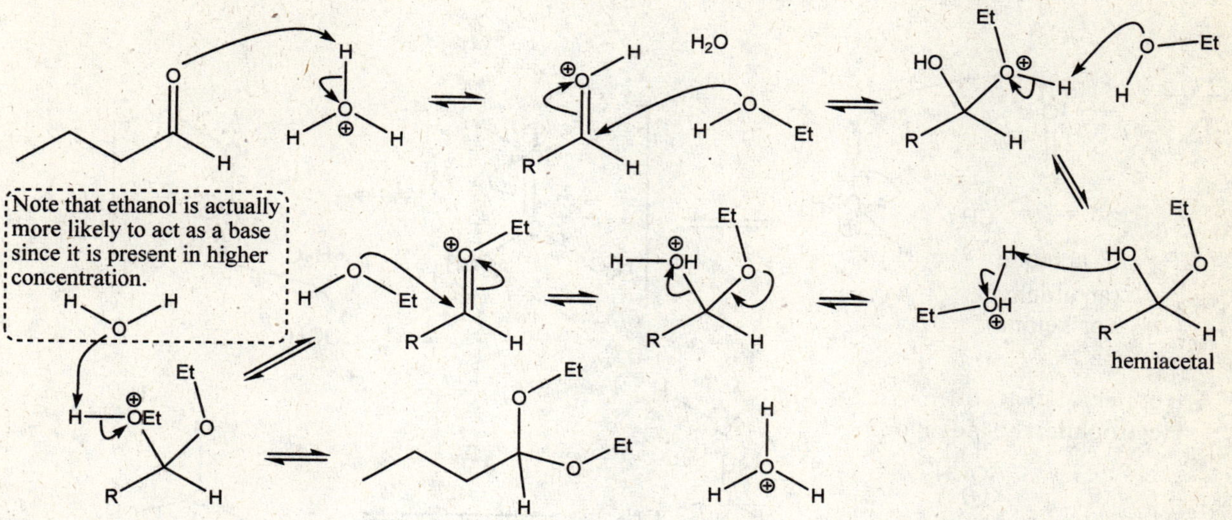

Note that ethanol is actually more likely to act as a base since it is present in higher concentration.

a. See above.

b. To push the above reaction toward a high yield of this acetal you must use an excess of ethanol and remove water as it forms.

21.

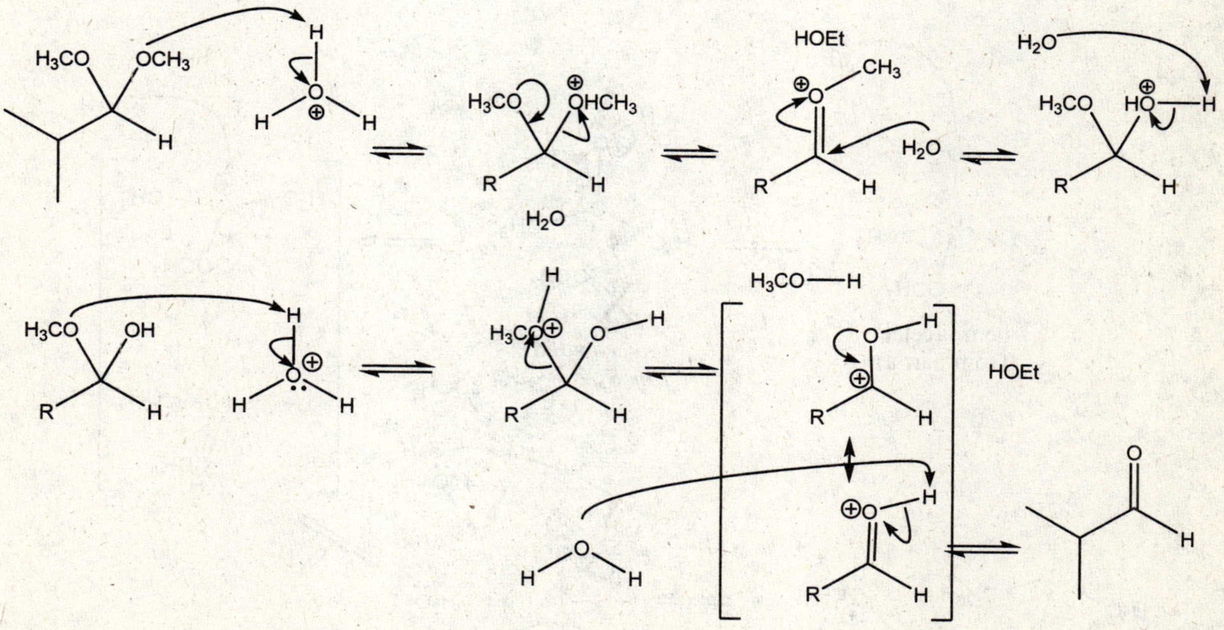

22.

- hemiacetal: $CH_3CH_2-C(OH)(OEt)(H)$
- acetal: $Ph-C(OCH_3)(OCH_3)(H)$ (with H$_3$CO and OCH$_3$)
- hemiketal: $H_3C-C(OH)(OCH_3)(CH_2CH_3)$
- ketal: $H_3C-C(OEt)(OEt)(Ph)$

23. a) [Mechanism shown: acetone (an aldehyde or ketone) + protonated methanol → protonated intermediate (resonance structures) → hemi-acetal (unstable) after deprotonation by :OCH$_3$]

[Second stage: hemi-acetal (from part a) → protonation of OH → loss of water (X — too crowded for S$_N$2) → oxocarbenium ion (resonance structures) via S$_N$1 → attack by :OCH$_3$ → deprotonation → acetal (stable)]

Volume 2, ChemActivity 7: Addition-Elimination

Continued from previous page

b) [reaction mechanism scheme]

c) [reaction mechanism scheme]

Probably a solvent molecule would act as base instead of chloride but the solvent is not specified.

cyclic hemi-acetal (stable)

Volume 2, ChemActivity 7: Addition-Elimination

Continued from previous page

d) acetone (an aldehyde or ketone)

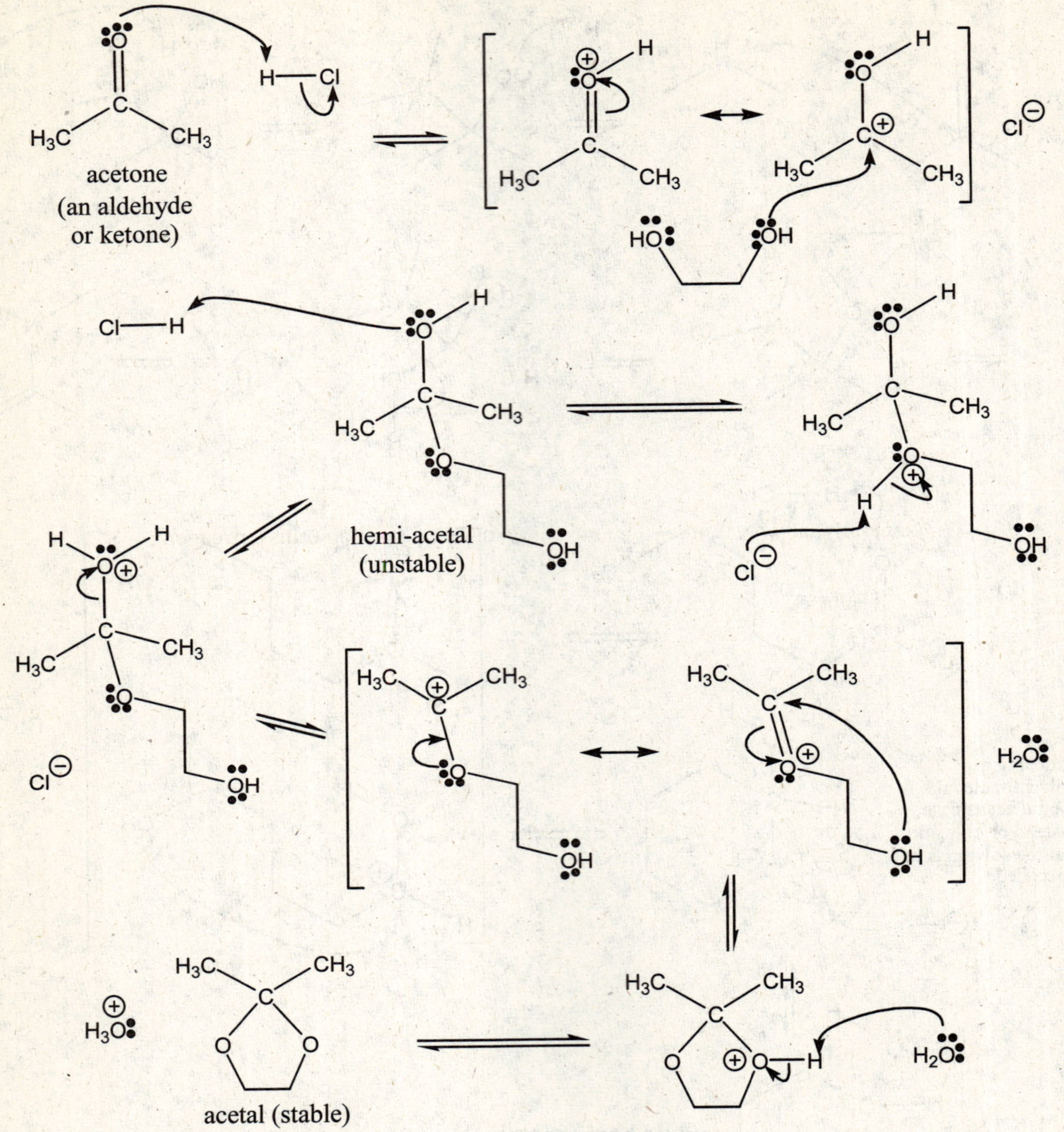

hemi-acetal (unstable)

acetal (stable)

24. In this question, two different ways to show a proton transfer are shown.

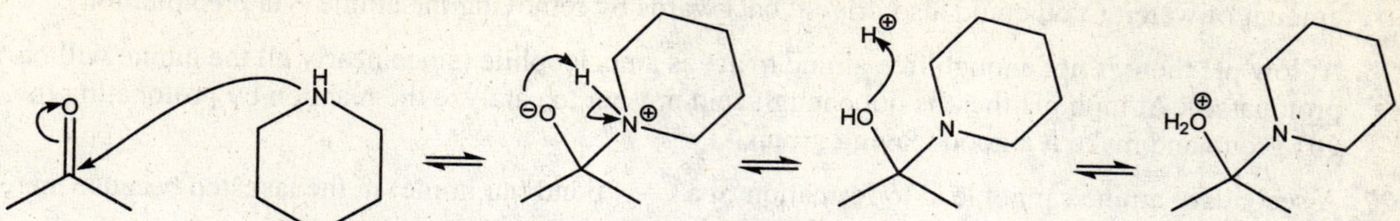

Intramolecular proton transfer via a four member ring transition state is much less likely than a solvent or amine molecule acting as a proton transfer reagent in two steps, as shown in the next mechanism.

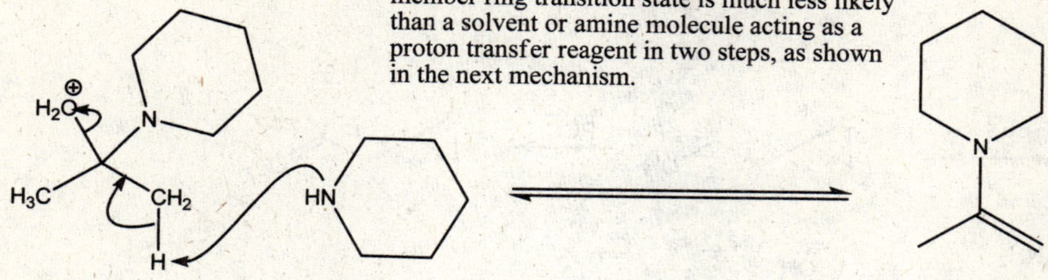

Generally, any solvent with a lone pair can act as a proton transfer reagent.

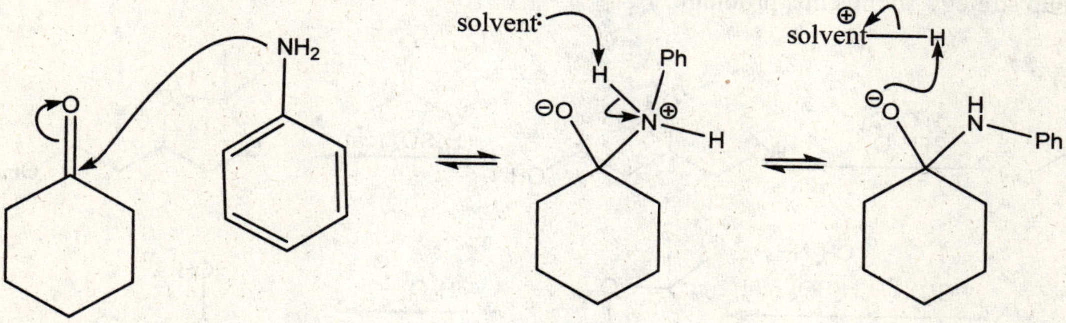

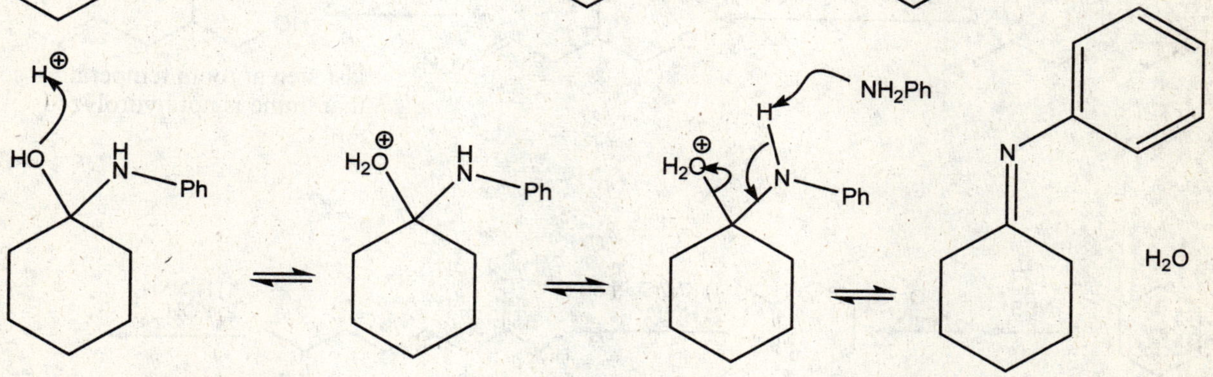

For both the reactions above, many students propose that the first step is protonation of the carbonyl with the acid catalyst. This gets you to the right product and does not violate any key rules; however, based on relative pK_a values, the pathway shown above is more likely. Think of it this way. There is no free "H^+" in solution. It is almost all bound to amine. The pK_a of the conjugate acid of a carbonyl oxygen is below zero. This means it is very unlikely to pull an H^+ off an ammonium ion (which requires 9 pK_a units of energy). In contrast, after the amine binds to the carbonyl, the result is an anionic oxygen, which is much better able to compete with the ammonium ion for an H^+.

25. Because water is a product of the reaction, you could drive it back toward ketone by adding a large amount of water. (You could also drive it backwards by removing the amine (via precipitation).

26. At low pH there is not enough free amine to act as a nucleophile (since nearly all the amine will be protonated). At high pH there is not enough acid present to catalyze the reaction by protonating the OH group and make it a good leaving group.

27. A secondary amine cannot lead to formation of a C=N bond (an imine) in the last step because there is no H on the nitrogen to be eliminated by the base.

28.

In the last step, above, the amine may also open the epoxide. An alternate answer, involving a protecting group strategy, avoids this problem.

last step at room temperature so that imine is not hydrolyzed

Continued on the next page

The following is an alternate to the previous answer that uses a protecting groups strategy.

29.

30.

31.

32. a. Yes. The reaction profile for base catalyzed hydration shows two steps, and the mechanism above is two steps.

b. In the reaction of an aldehyde with a Grignard the starting materials are much higher in energy. (A negatively charged carbon is higher in PE than a negatively charged oxygen, due to electronegativity effects, and the product is resonance stabilized.) In contrast, in the hydration reaction (either acid or base catalyzed), the reactants and products are very similar in energy. A good way to determine if two molecules are similar in energy is to examine where the negative charge (if any) resides. In the case of base catalyzed hydration, the negative charge resides on oxygen in both cases (and neither molecule is resonance stabilized). This indicates they are similar in energy. It is usually a good assumption that two neutral molecules with similar functional groups will be similar in potential energy. Based on this assumption, the acid-catalyzed hydration reaction is also thermoneutral.

c. A reaction that is neither uphill nor downhill (thermoneutral) is considered "reversible" because the rate of the reverse reaction is similar to the rate of the forward reaction. That is, once a product forms, there is nothing (other than equilibrium considerations, i.e. Le Châtelier's rule) preventing it from reverting back to starting material. In contrast, the Grignard reaction is downhill to the right. Once product forms it is very difficult, energetically, for it to go back toward the starting material. That is, the reverse reaction is much much slower (higher activation energy/uphill) than the forward reaction. Such reactions are considered "irreversible."

33.

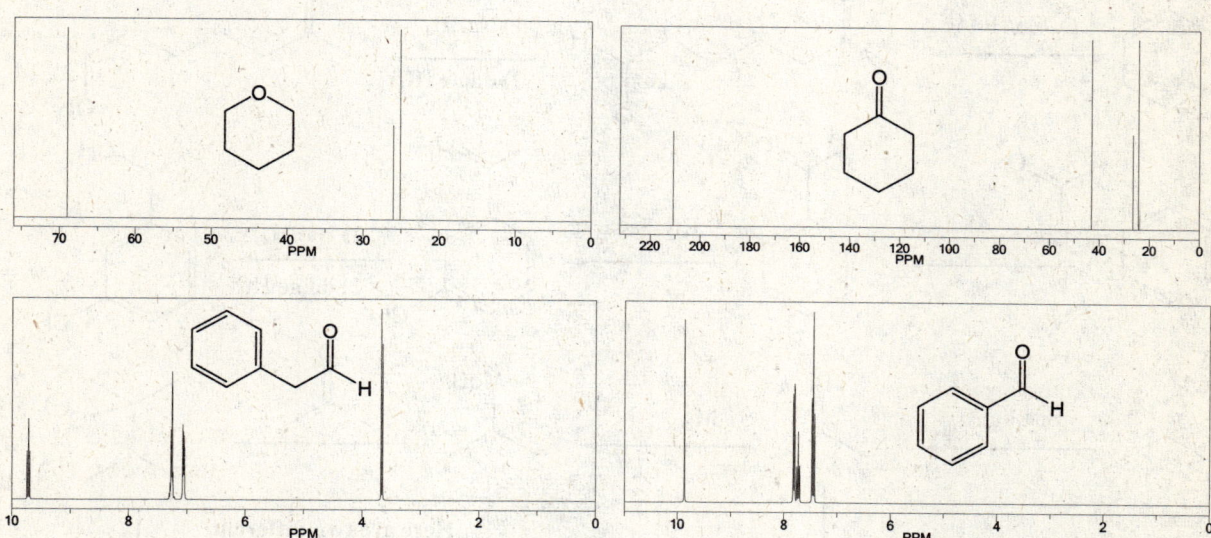

34. Oxidation of 1-butanol to butanal can be accomplished using pyridinium chlorochromate (PCC).

35.

36.

37. Because aldehydes are more reactive than ketones, the cyanide nucleophile would be more likely to form a cyanohydrin with the aldehyde.

38.
    a. A shorter synthesis consisting of treatment of the starting material with sulfuric acid/nitric acid will yield the *para* product, not ortho.

    b. The protecting group in this synthesis (an SO_3H group) prevents substitution of the nitro group at the para position, forcing to go to the more hindered ortho position.

39. Design a synthesis of each target from the starting material given.

ChemActivity 8: Carboxylic Acids & Derivatives

Extend Your Understanding Questions (to do in or out of class)

8. This reaction does NOT produce the desired product because the amine is much more basic than the carbonyl oxygen. This means it is impossible to have enough acid in solution to protonate the carbonyl oxygen without protonating nearly all of the amine molecules. That is, in the presence of excess acid, nearly all of the amine will be in the protonated (non-nucleophilic) form.

9. Reason 2 in Memorization Task 8.4 explains the failure of the reaction at the top of the page.

10.

11. a.

    b. The conjugate base of acetic acid is formed preferentially over the nucleophilic addition product above because the result of the acid-base reaction is much lower in potential energy, as demonstrated by the fact that it is resonance stabilized.

12. Addition of a large amount of water and/or removal of the alcohol would allow you to reverse this reaction and generate a carboxylic acid from an ester as shown in Synthetic Transformation 8.3. Note that this can be achieved with ease when the alcohol boils at a temperature below 100 degrees since this allows you to distill off the alcohol as it forms.

13. Mechanism of Synthetic Transformation 8.3.

14. Mechanism of Synthetic Transformation 8.4.

15. Mechanisms of Synthetic Transformation 8.5, acid- and base-catalyzed trans-esterification.

    Acid-catalyzed trans-esterification

    Base-catalyzed trans-esterification

16. Mechanisms of Synthetic Transformation 8.6, acid- and base-catalyzed amide hydrolysis.

Acid-catalyzed amide hydrolysis

Base-catalyzed amide hydrolysis

17.

18. Synthetic Transformations 8.4 through 8.6, which do work in basic conditions, are different because they do not begin with a carboxylic acid. The acidic H on a carboxylic acid is removed whenever it is mixed with a basic nucleophile. The removal of this H generates a carboxylate ion, which, due to its negative charge, is a much much worse electrophile than an ordinary carbonyl.

Confirm Your Understanding Questions (to do in class)

19. The amide (on the far right) in Model 1 has the most double-bond character in the bond marked in the top structure with an arrow => based on the argument that N is the best pi electron donor represented. This allows it to donate electron density toward the carbonyl carbon, forming a partial pi bond between the N and the carbonyl carbon.

20. In the carbon NMR spectrum of *N,N*-dimethylformamide two different methyl peaks (at 36 ppm and 32 ppm) are found due to the partial double bond between N and the carbonyl carbon. The partial double bond restricts rotation of this bond at room temperature. The result is that the NMR detects one methyl group that is *cis* to the oxygen and one methyl group that is *trans* to the oxygen.

21. Yes, at elevated temperatures there is enough thermal energy to overcome the partial pi bond. Free rotation about the N—C bond of *N,N*-dimethylformamide means that there is no longer a difference between the two methyl groups.

22.

23.

24.

25.

26.

27. The following is an S$_N$2 reaction.

ChemActivity 9: Acid Halides and Anhydrides

Extend Your Understanding Questions (to do in or out of class)

10. Esters are <u>less reactive</u> than acid anhydrides or acid halides in addition-elimination reactions because an OR group is a better pi donor than a halogen or acyl group. This means the carbonyl carbon of an acid halide or anhydride is more electrophilic. Additionally, a halogen or acyl group is a much better leaving group than an OR group.

11.

12. **f** (f with an ester is the same as d/e), **g**, and **h/i (LiAlH₄ only)** are key reactions that work with esters. (In sufficiently basic or acidic conditions, **a, b,** and **c** also work with an ester.)

13. The functional group produced by reaction a = carboxylic acid, b = amide, c = ester, d = ketone, e/g = alcohol (tertiary), f = ketone, h = aldehyde, i = alcohol (primary). The key reactant in reaction a = water, b = amine, c = alcohol, d-g = Grignard or lithium reagent, h/i = hydride.

Confirm Your Understanding Questions (to do at home)

14. a. 2,6-dimethylpyridine is better than pyridine as a base in this reaction because the former is less nucleophilic than the latter, and is much less likely to make a bond to the carbonyl carbon of the acid halide.

  b. No auxiliary base is needed in this question because the amine itself acts as a base (as shown below). The other product not shown is the ammonium ion of ethyl amine.

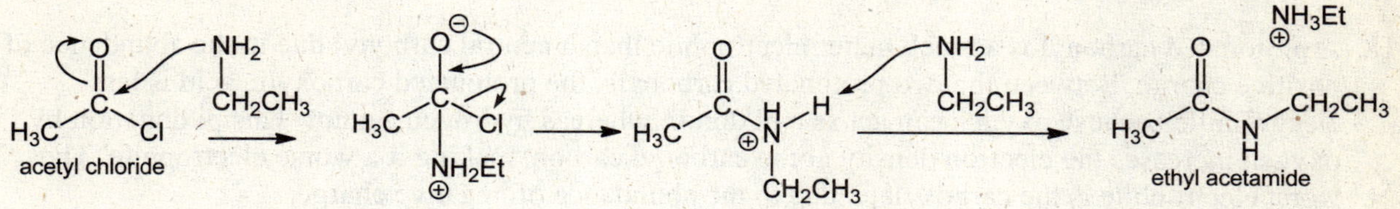

15.

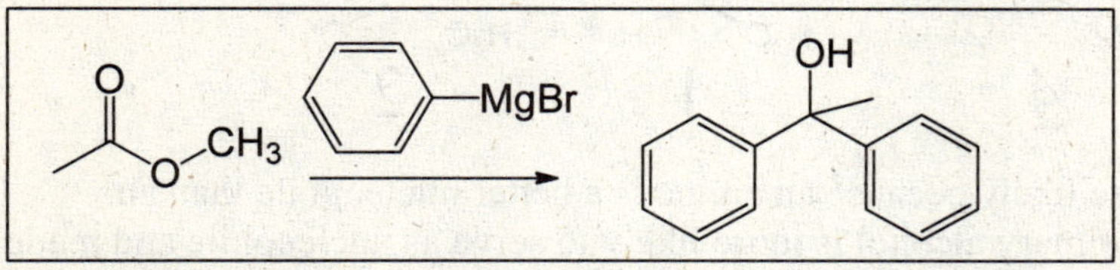

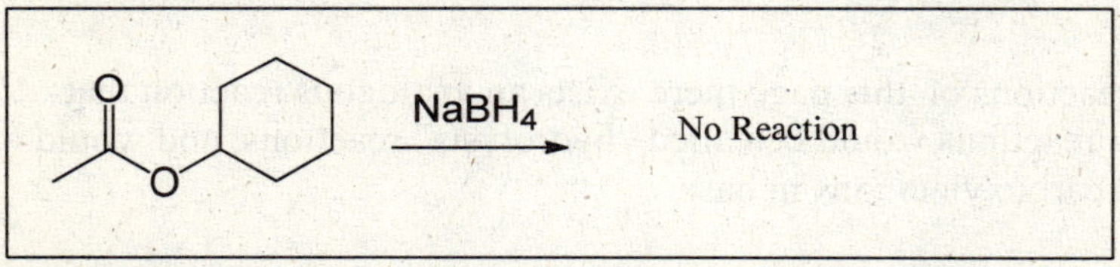

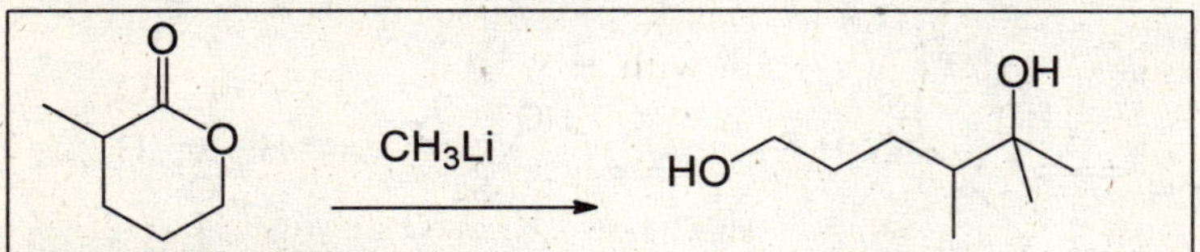

example of a lactone

16.

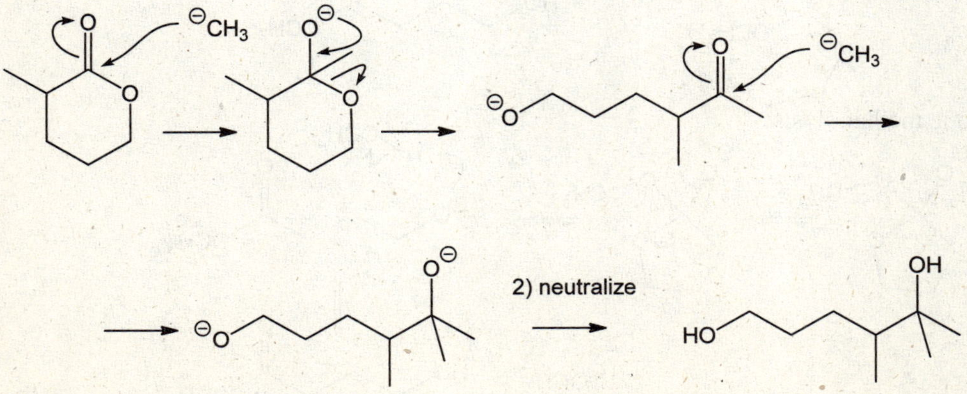

17. a. The first, forth, and fifth anhydrides are symmetrical, the third and fourth are asymmetrical.
    b. use of an asymmetrical acid anhydride usually gives rise to two different products because either carbonyl is subject to reaction with the nucleophile, and either acyl group can leave.
    c. *See below.*

18. A protonated carbonyl is a much better electrophile than a neutral carbonyl due to the abundance of positive charge. Between the two protonated carbonyls, the protonated carboxylic acid is less electrophilic because oxygen can act as a pi donor, whereas hydrogen cannot. This pi donation by oxygen increases the electron density at the carbonyl carbon, making it a worse electrophile. The worst electrophile is the carboxylate, due to the abundance of negative charge.

    3   4   1   2

19. In the first reaction, the amide is more likely because an amine is a better nucleophile than an alcohol. In the second reaction, the primary alcohol is more likely to serve as nucleophile and made a bond to the carbonyl due to sterics.

20. Note that for the four methanolysis reactions on this page there exists an analogous reaction that uses water instead of methanol. Such reactions would be called "hydrolysis" reactions, and would result in carboxylic acids in acid and carboxylate ions in base.

    Acid-catalyzed lactone methanolysis

    Base-catalyzed lactone methanolysis

Continued on next page.

Acid-catalyzed lactam methanolysis

Base-catalyzed lactam methanolysis

Acid-catalyzed lactone aminolysis

Base-catalyzed lactone aminolysis

Continued on next page.

Acid-catalyzed lactam aminolysis

Base-catalyzed lactam aminolysis

21. *As noted for each, reference the appropriate mechanism from the previous question.*

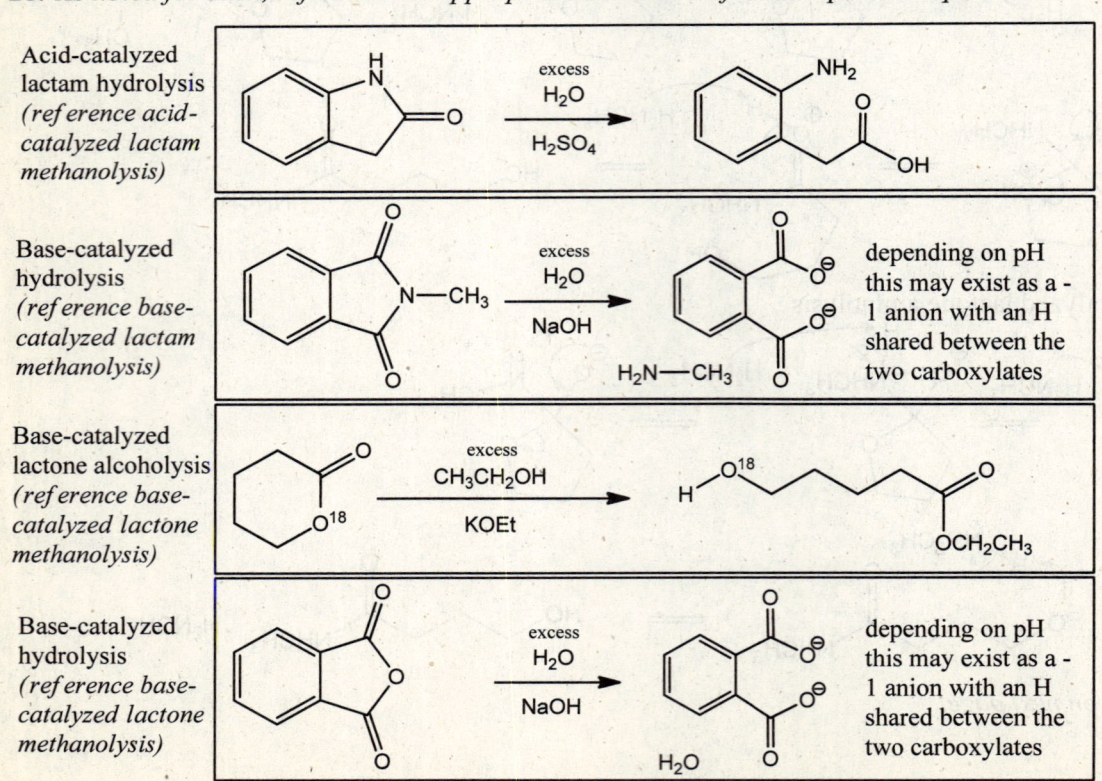

22. The ester and carboxylic acid from a given mechanism are boxed together.

23. a. a halide (Cl⁻ or Br⁻)
     b. a carboxylate (RCO$_2$⁻)
     c. Mechanism of reaction of an ester with a Grignard reagent (followed by acid).

24.

25.

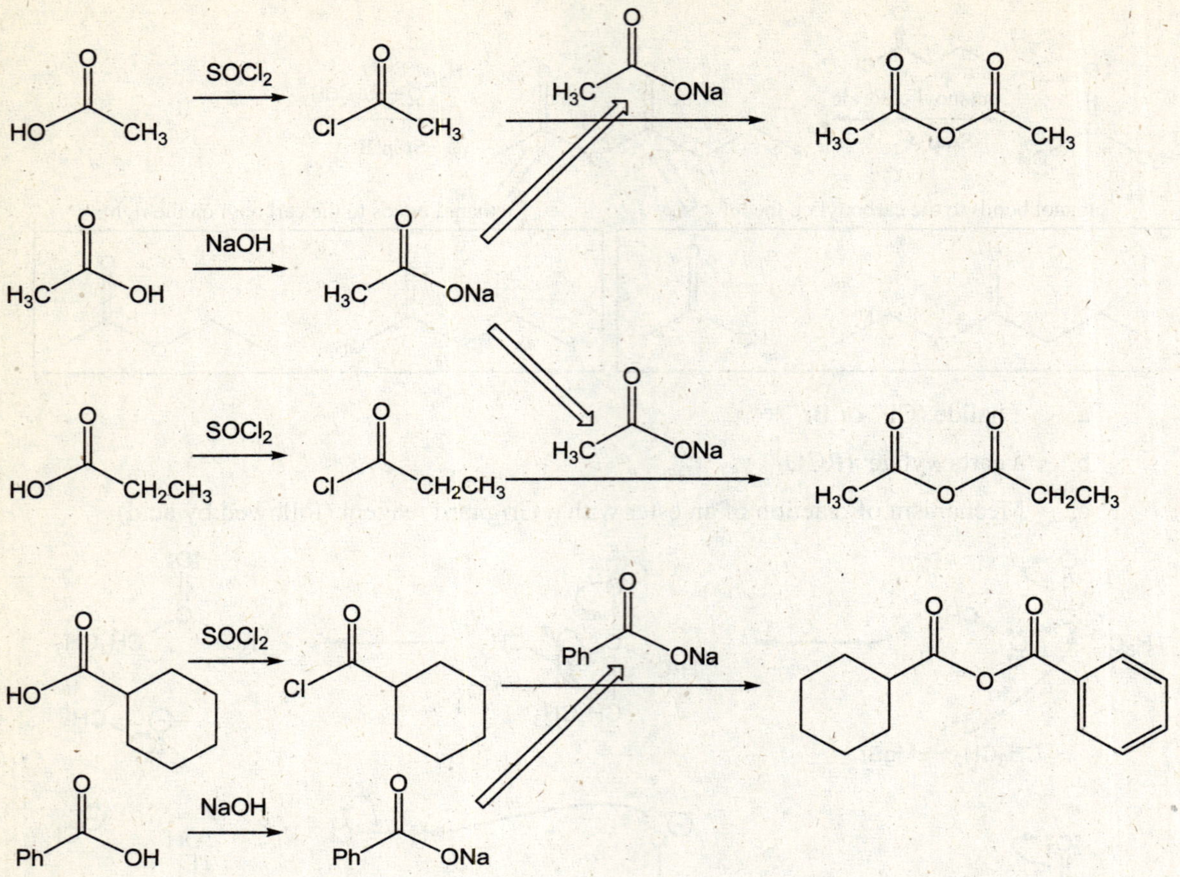

26. The other product in this case is the carboxylic acid shown below. While this acid can be recovered and used at the start of synthesis to make more acid anhydride, each of these sets is associated with some loss, and eventually, the final time you run the synthesis, there will be some chiral acid produced that will not be incorporated into the product. This means there will be some waste of the expensive chiral alkyl group, R*. For these reasons it may be better to choose a synthesis involving an acid chloride, even though some acid chloride reactions proceed so quickly that other side-products can be formed.

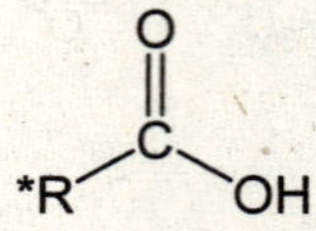

27. This asymmetrical anhydride reaction gives a much higher yield of ethyl acetate than ethyl 2,2-dimethylpropionate because the ethanol preferentially bonds to the less hindered carbonyl.

28.

29. Design an efficient synthesis of each of the following target molecules.

ChemActivity 10: Enolate and Enol Nucleophiles

Extend Your Understanding Questions (to do in or out of class)

10. a. Alcoxide (RO⁻) is NOT found in an acidic solution (*except in trace amounts*) because there is so much H⁺ in solution that it would protonate most alcoxide as it formed. Another way of saying this is that you cannot have a strong base in acidic solution or vice versa.
    b. The best nucleophile available for reaction in a **basic** solution of aldehyde is the enolate of that aldehyde, but the best nucleophile available for reaction in an **acidic** solution of aldehyde is the enol of that aldehyde.
    c. RO is a better nucleophile⁻ than ROH.
    d. enolate is a better nucleophile than enol.
    e. Yes.

11. Most of the propanal in Model 4 can be converted to 2-bromopropanal even though, at any given moment, less than 1% of the propanal in solution is in the enol form. This is because, as enol is converted into 2-bromoporpanal, (according to Le Chatalier's principle) more keto aldehyde is converted into the nucleophilic enol form.

12. The fourth, fifth, and sixth molecules in Model 5 are β-keto carbonyl compounds.

13. a. *See below.*
    b. An aldehyde is more acidic than a comparable ketone.
    c. A ketone is more acidic than a comparable ester.

[Structures labeled 24, 19, 17, 13, 11, 9, 5]

14. [Reaction scheme showing β-keto ketone with ⁻OH forming resonance-stabilized enolate]

15. The β-keto ketone above (with a pK_a of about 9) is much more acidic than an ordinary ketone (with a pK_a of about 19) because the conjugate base is resonance stabilized, and therefore easier to generate from the β-keto ketone. That is to say, it takes less energy to generate the conjugate base o of the β-keto ketone.

16.

17.

Confirm Your Understanding Questions (to do at home)

18. One explanation for why it is much easier to remove an H from a carboxylic acid than from an aldehyde is based on a comparison of the two conjugate bases. The conjugate base of a carboxylic acid (a carboxylate) has two resonance forms, both with a negative charge on oxygen. In comparison the conjugate base of an aldehyde (an enolate) has one resonance form with a negative charge on oxygen and one resonance form with a negative charge on carbon. Since a negative charge on oxygen is much more favorable than a negative charge on carbon, we can predict that the carboxylate is lower in potential energy than the enolate. The lower energy product requires less energy to make, therefore it is easier to remove an H from a carboxylic acid than an aldehyde.

19.

20. Each of the examples is called a **β-keto** aldehyde or ketone because there is a carbonyl group on the carbon beta to a carbonyl. More information on proper use of the term "keto" as part of a name can be found in Nomenclature Worksheet 4.

21.
a.
b.
c.

[resonance structures of a β-keto aldehyde enolate shown] ↔ [resonance structures shown]   not resonance stabilized

d. [four structures showing alpha hydrogens marked on β-dicarbonyl compounds and cyclohexanone derivatives]

22. LDA is used in Synthetic Transformation 10.1 because the pK_a of an ordinary ketone is around 20. (Alpha alkylation is best done with a base that is strong enough to convert nearly all of the ketone into enolate since this prevents competition from aldol type reactions we will be studying in the next chapter.) Making a bond to LDA releases about 35 pK_a units of energy so this is plenty to remove the H alpha to any carbonyl. In contrast, an alkoxide base only releases about 16 pK_a units of energy when it makes a bond to an H. This is not enough to remove an ordinary aldehyde or ketone alpha H (which requires 20 pK_a units of energy), but is enough to remove the alpha H of a β-keto compound (which requires only about 9-13 pK_a units of energy).

23. Students often confuse these two structures because they both have carbonyl groups separated by a single atom. However, a β-keto carbonyl compound (such as a malonic ester) has a carbon separating the two carbonyl groups, whereas an acid anhydride has an oxygen separating the two carbonyls. This makes a big difference in terms of the reactivity of the two molecules. A β-keto carbonyl compound often reacts by giving up an alpha H, whereas an acid anhydride is valuable because it is an excellent electrophile (similar to an acid chloride).

24. Step 2 is faster. One explanation is that bromine serves as an electron withdrawing group. This tends to polarize the carbon-H_{alpha} bond, partially breaking it and making this H more acidic.

25. Yes, the answer above is consistent with the fact that base-promoted α-halogenations are very difficult to control and usually result in all α-H's being replaced with halogen atoms. Once the first alpha H is replaced with a halogen, subsequent alpha H's are easier to remove. Since removal of the alpha H is the rate limiting step in each case, the reaction goes faster and faster and only stops when all alpha H's have been replaced or all halogens have been used up.

26.

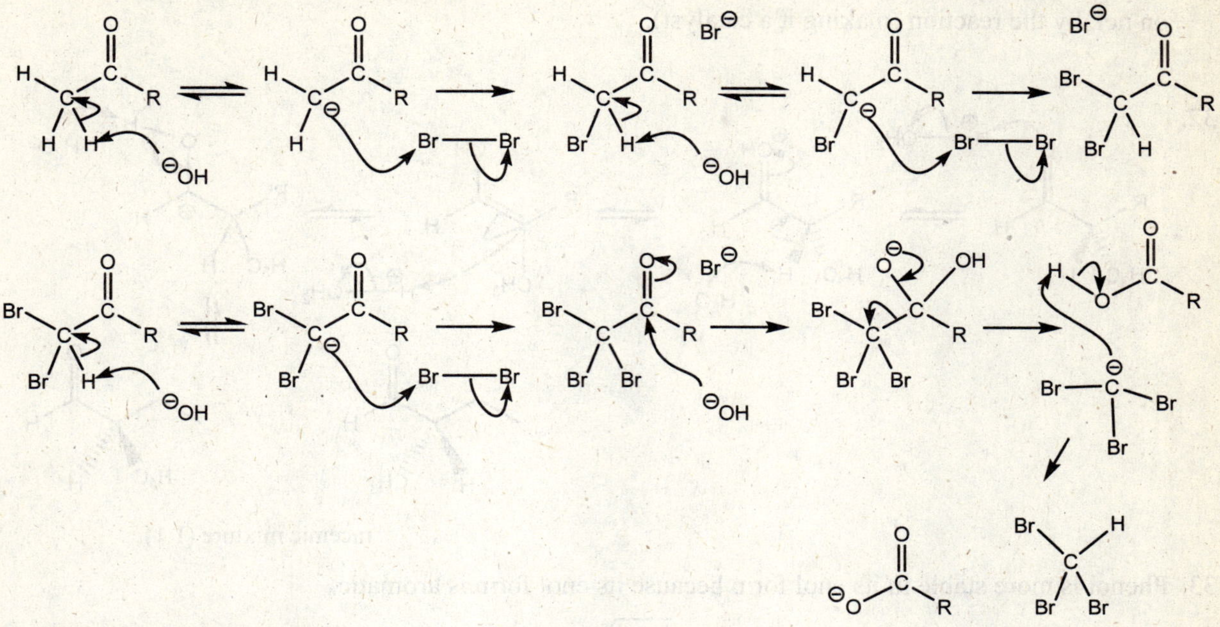

27. Keto-enol tautomerization

28. neither (rates are the same)

29.

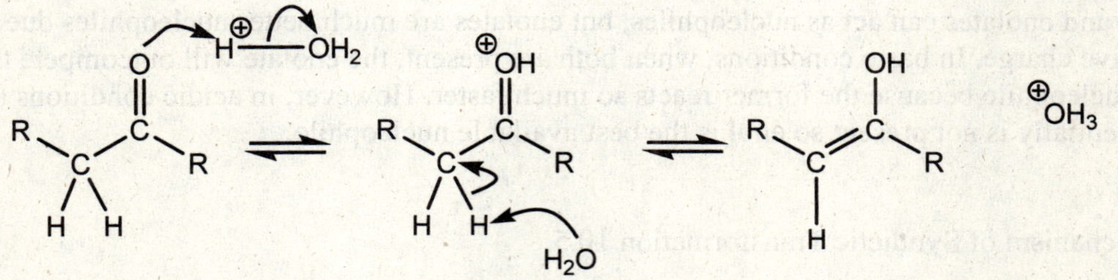

30.

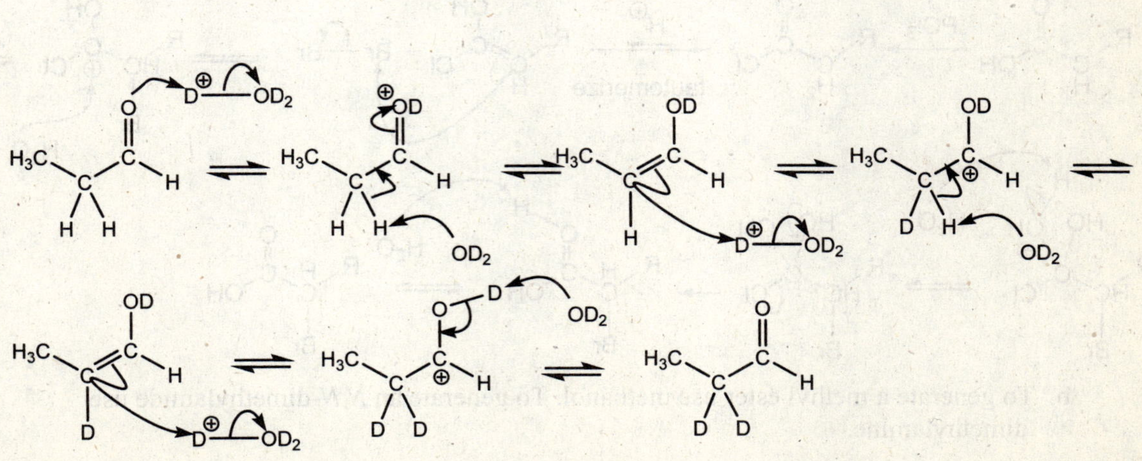

31. The reaction in Model 3 is called base-"promoted" halogenation because one full molar equivalent of base is needed, while the reaction in Model 4 is called acid-"*catalyzed*" halogenation because only a relatively small amount of acid is needed, and this acid is neither consumed or produced, on net, by the reaction (making it a catalyst).

32.

racemic mixture (1:1)

33. Phenol is more stable in its enol form because its enol form is aromatic.

34. Both enols and enolates can act as nucleophiles, but enolates are much better nucleophiles due to their negative charge. In basic conditions, when both are present, the enolate will outcompete the enol as a nucleophile because the former reacts so much faster. However, in acidic conditions the enolate essentially is not present so enol is the best available nucleophile.

35.   a. Mechanism of Synthetic Transformation 10.5.

   b. To generate a methyl ester use methanol. To generate an *N,N*-dimethylamide use dimethylamine.

36. Note that alanine does not exist at any pH in the form shown in the question. At neutral pH it is a zwitterion, and at acidic pH it is in the form shown below.

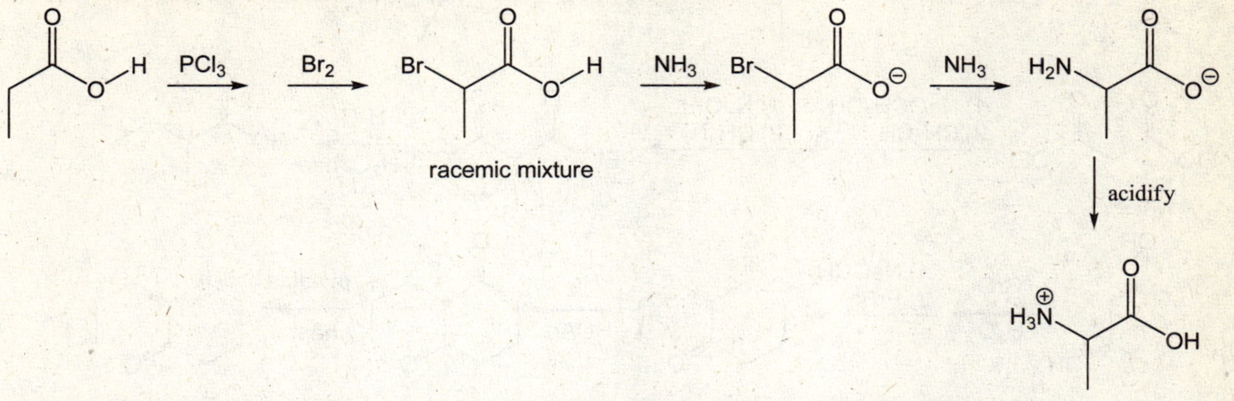

37. The product of the reaction above is racemic alanine, rather than the biologically active S enantiomer, L-alanine because in this procedure (as in most ordinary laboratory procedures) there is no force driving the reaction toward one enantiomer or the other. To selectively produce one enantiomer a chiral catalyst must be used. In biological systems L-alanine is selectively generated by a chiral enzyme catalyst. In the laboratory great effort and expense goes into making and using synthetic chiral catalysts. Such enantioselective catalysis is perhaps the greatest challenge of modern synthetic organic chemistry.

38. Design a synthesis of each of the following target molecules from the starting material given.

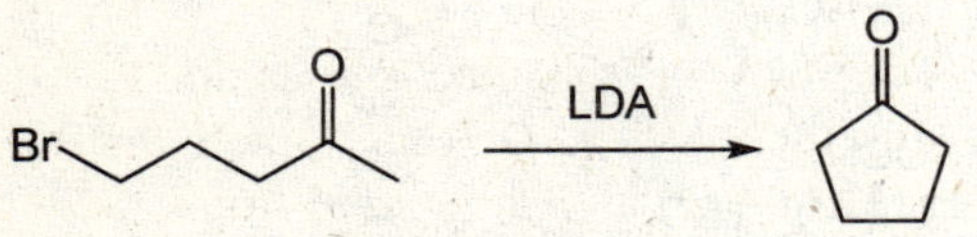

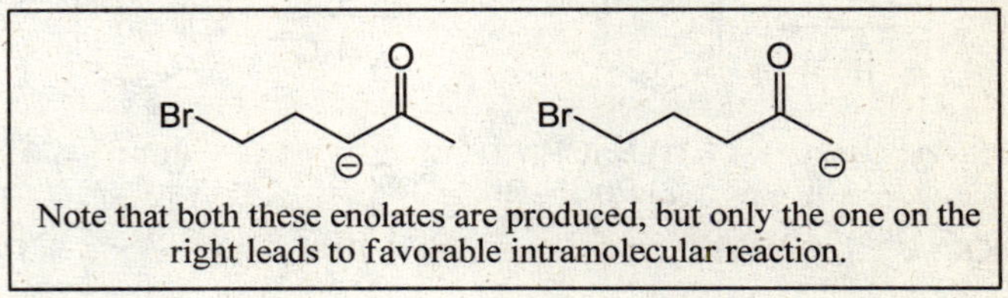

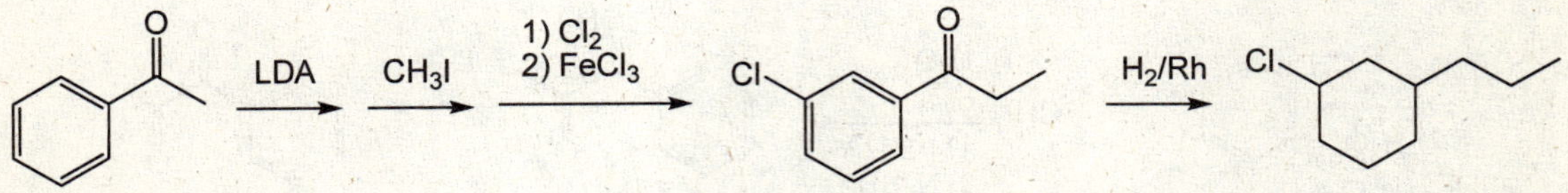

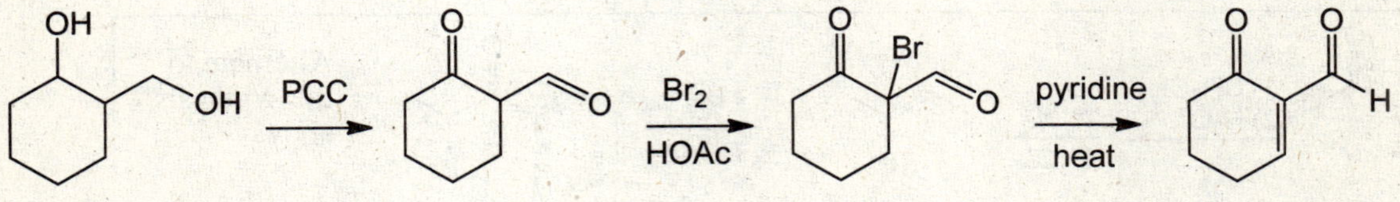

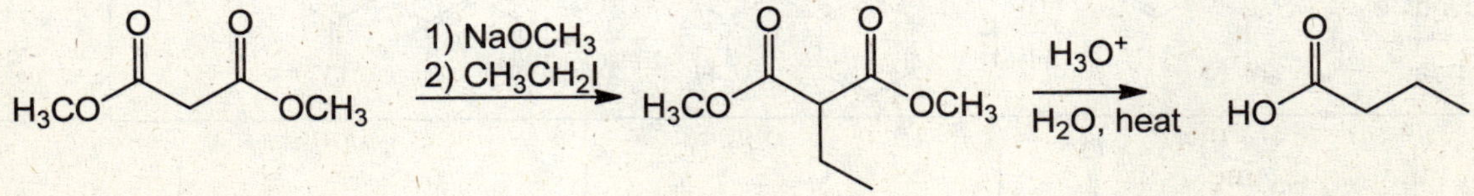

Continued on next page.

Volume 2, ChemActivity 10: Enolate and Enol Nucleophiles

ChemActivity 11: Aldol Reactions

Extend Your Understanding Questions (to do in or out of class)

12. In general an aldehyde is a better electrophile than a comparable ketone.

13. Yes.

14. a. The aldol reaction in Model 1 is approximately thermoneutral (neither up nor downhill).
    b. The ketol reaction in Model 4 is uphill.

15. Formation of the two crossed out products below is unfavorable because, to form them, the ketone must act as the electrophile. Since an aldehyde is available, it is more likely to act as electrophile.

electrophile ⟹	acetaldehyde	acetaldehyde	acetone	acetone
nucleophile ⟹	acetone	acetaldehyde	acetone	acetaldehyde

16. Yes.

17. In the last two reactions, acetaldehyde is most likely to act as both nucleophile and electrophile.

18. Yes

Confirm Your Understanding Questions (to do at home)

19. An aldehyde without an alpha H cannot undergo an aldol reaction.

    An aldehyde with an alpha H can undergo an aldol reaction.

20.

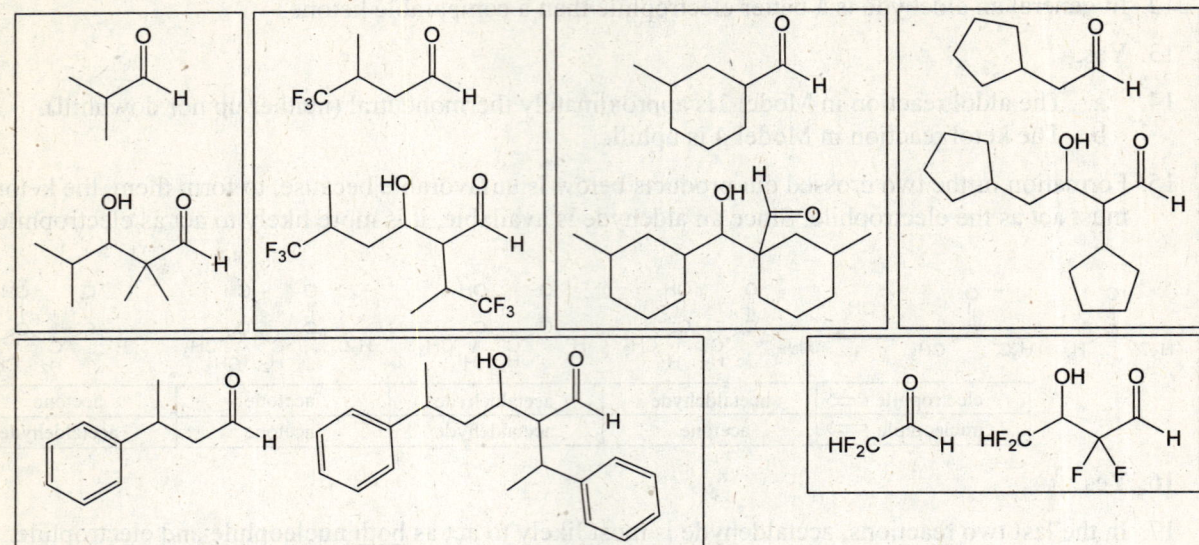

21. a.

    b. hydrate formation

    c. Relative amounts of two species in equilibrium at equilibrium is a function of their relative potential energies. Because the aldol product is lower in PE than the hydrate, it is expected that the aldol dominate at equilibrium.

22.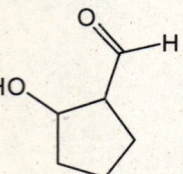

    a. The <u>intra</u>-molecular product is much more likely to form because two functional groups that are tethered together are more likely to collide and react with one another. This is especially true when the reaction yields a favorable ring size (5, 6 & 7 are most common).

    b. At very high concentrations of di-aldehyde you would expect to see a some of the intermolecular product, and even some chains (polymers).

23.

In the product above, left, the alpha C of the 6-aldehyde acted as nucleophile and the 1-aldehyde acted as electrophile. The opposite is true for the product above, right.

24.

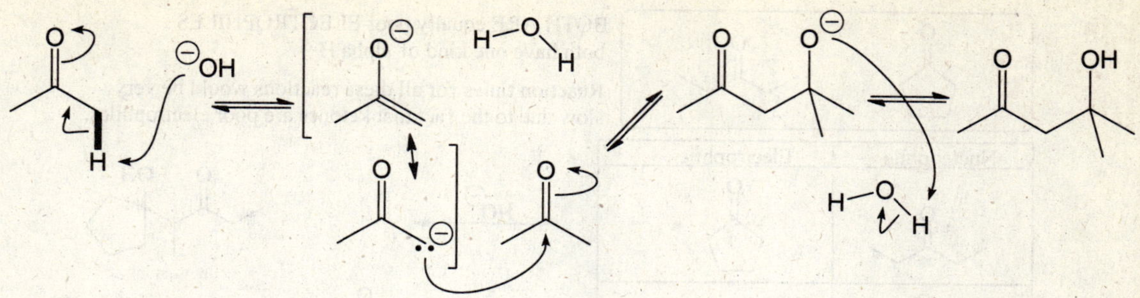

25.

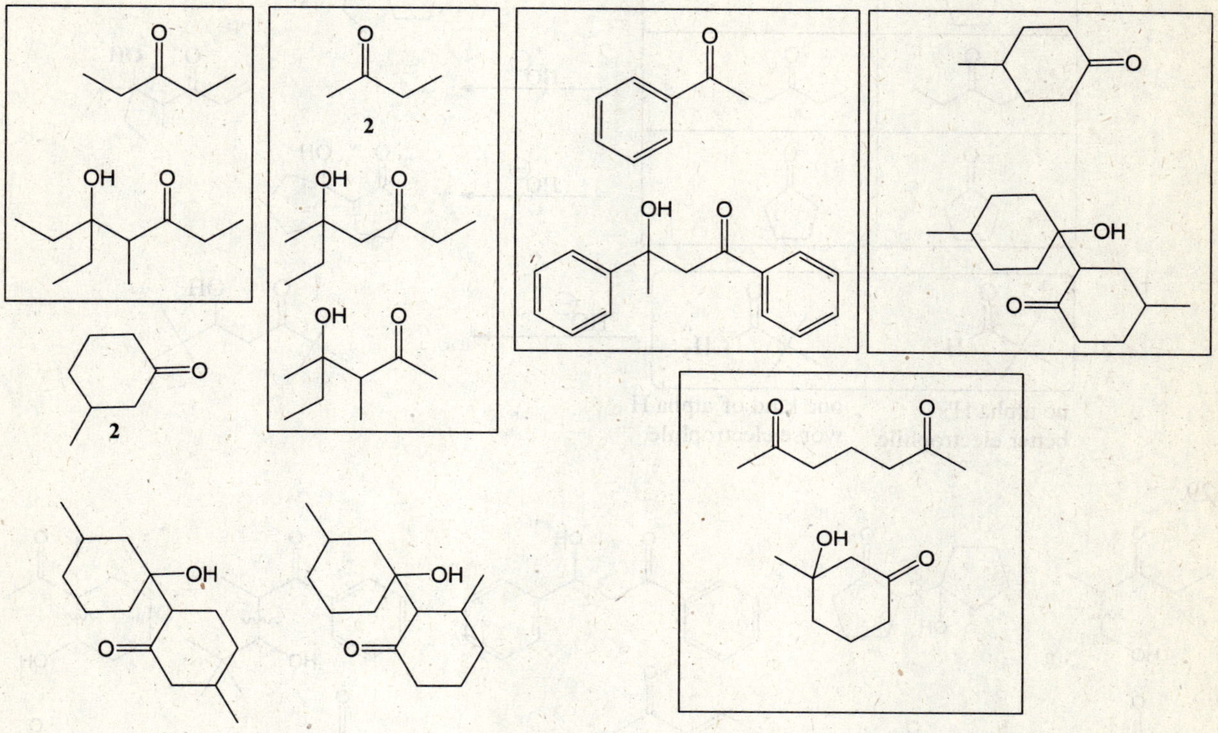

26. Any ketone with two different kinds of alpha H's can give rise to two different aldol/ketol self-condensation products

27. Benzaldehyde cannot act as a nucleophile in an aldol reaction because it has no alpha H's.

a. Benzaldehyde is a better electrophile than acetophenone.

b.

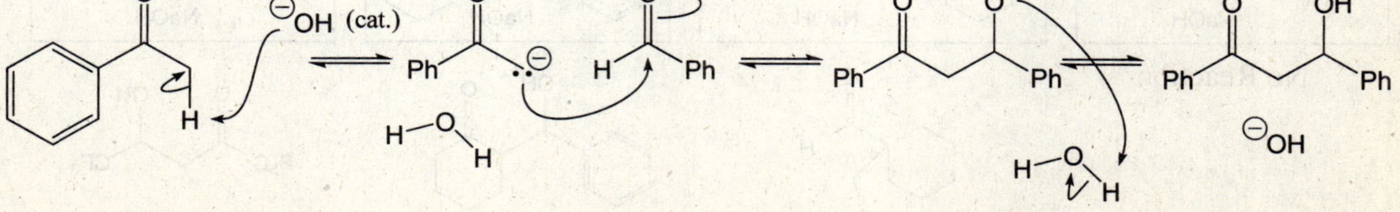

28. a.

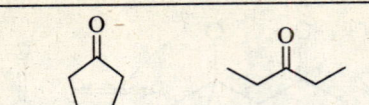

BOTH ARE equally poor ELECTROPHILES
both have one kind of alpha H

Reaction times for all these reactions would be very slow due to the fact that ketones are poor electrophiles.

b.

no alpha H's
better electrophile

one kind of alpha H
worse electrophile

29.

30. Draw the most likely product for each set of reactants, or write NO REACTION.

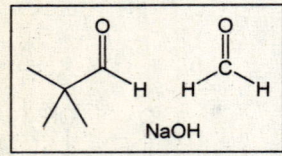

No Reaction

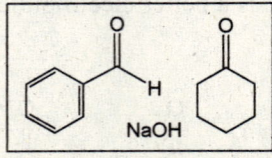

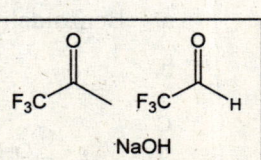

31. The following are called β-keto carbonyl compounds because there is a carbonyl group (sometimes called a "keto" group) on the carbon beta to the carbonyl.

   a.
   b.

   c.   The compounds above are more acidic than ordinary aldehydes or ketones?
   d.   Yes.
   e.

ChemActivity 12: Aldol Condensations

Extend Your Understanding Questions (to do in or out of class)

9.

10. This 2nd order resonance structure emphasizes the nucleophilicity of an enol because it shows a lone pair and negative charge on the alpha carbon.

11.

12. Procedure 2.

13. The box on the left should be crossed out. In this box, acetaldehyde will react to form an aldol product even before benzaldehyde is added.

14. In the box on the right, no aldol reaction can occur *until* a drop of the second aldehyde has been added because benzaldehyde cannot function as the nucleophile in an aldol reaction (since it has no alpha hydrogens).

15. Product A is *unlikely* to form in the second procedure because, for this to happen, the enolate of acetaldehyde would have to collide with a molecule of acetaldehyde. The concentration of benzaldehyde is much higher than the concentration of acetaldehyde. This means that any molecule of acetaldehyde enolate is much more likely to collide with a molecule of benzaldehyde.

Confirm Your Understanding Questions (to do at home)

16. The reactants below would give a high yield of the mixed aldol product shown <u>only</u> when acetaldehyde is slowly added to the benzaldehyde + base. This technique is useful for any pair of carbonyl compounds in which one does not have an alpha H, and thus cannot undergo self-condensation, and the other can under self-condensation.

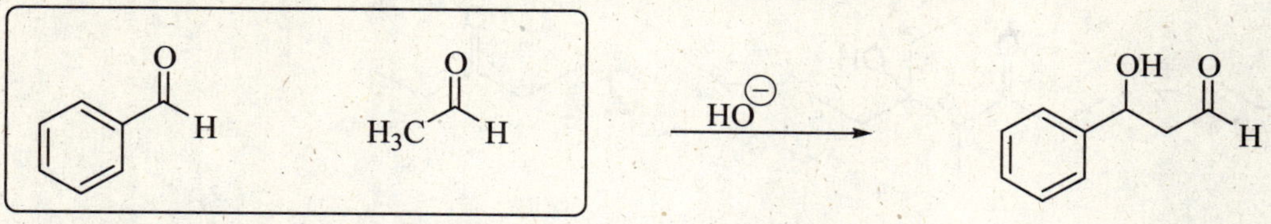

   [benzaldehyde] + [acetaldehyde H$_3$C-CHO] $\xrightarrow{HO^-}$ [Ph-CH(OH)-CH$_2$-CHO]

   no alpha H's   one kind of alpha H

   BOTH ARE GOOD ELECTROPHILES

17. An "α,β-unsaturated aldehyde" is one in which there is a pi bond between the alpha and beta carbons. The word unsaturated is used since the molecule does not have the maximum number of hydrogens (is not "saturated" with hydrogens).

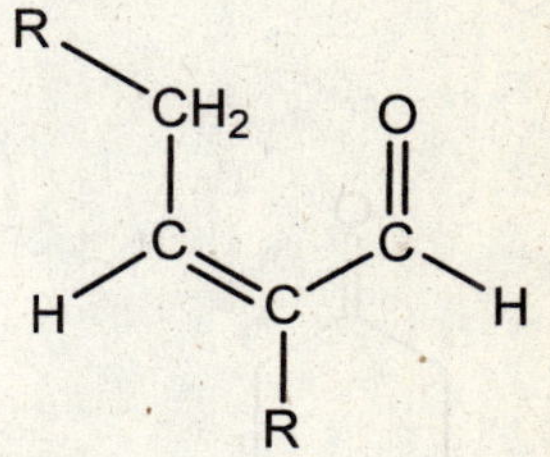

   a. H$_2$O

   b. If R = alkyl the molecule shown at right will be the Z stereoisomer.

18. Use retrosynthesis to determine starting materials that will give rise to each product.

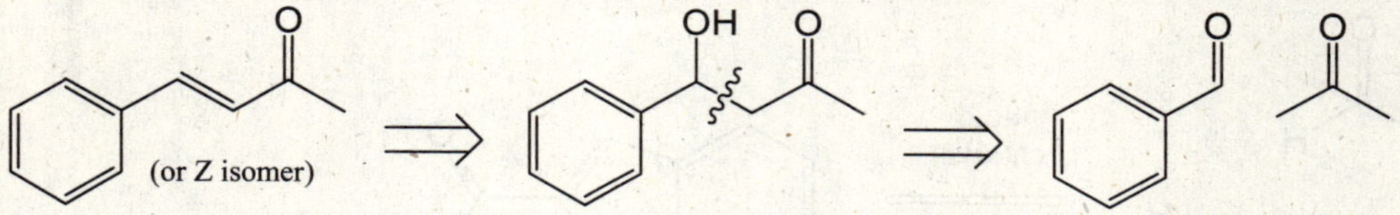

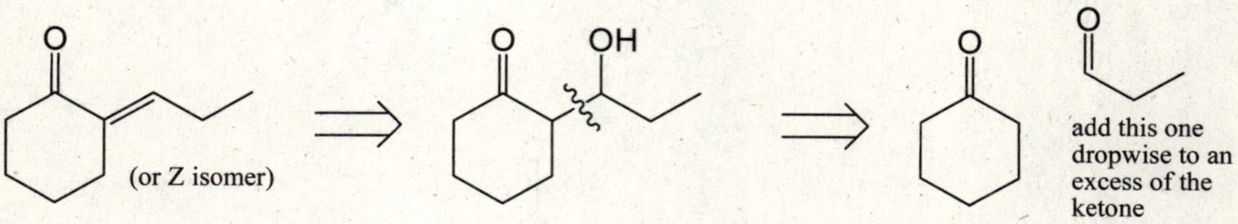

   [PhCH=CH-CO-CH$_3$] (or Z isomer) ⇒ [PhCH(OH)-CH$_2$-CO-CH$_3$] ⇒ [PhCHO] + [CH$_3$COCH$_3$]

   [cyclohexanone with =CHCH$_2$CH$_3$] (or Z isomer) ⇒ [cyclohexanone with -CH(OH)CH$_2$CH$_3$] ⇒ [cyclohexanone] + [propanal — add this one dropwise to an excess of the ketone]

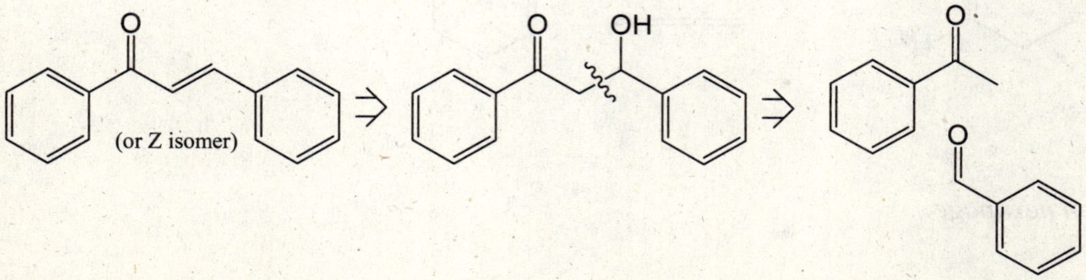

   [PhCO-CH=CH-Ph] (or Z isomer) ⇒ [PhCO-CH$_2$-CH(OH)-Ph] ⇒ [acetophenone] + [benzaldehyde]

19.

Continued on next page...

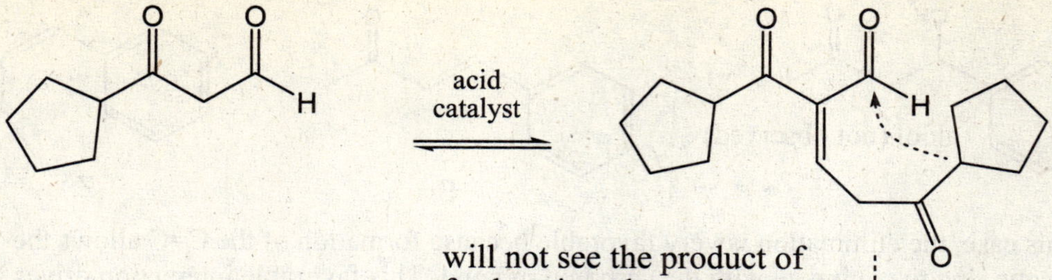

...will not see the product of further ring closure between these two carbons

...because elimination cannot occur since there are no H's beta to the alcohol

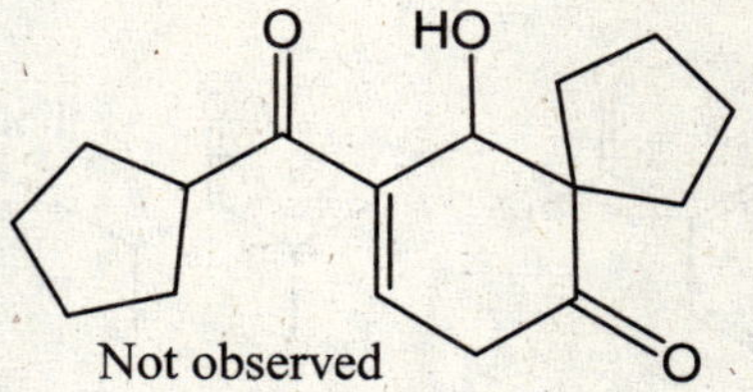

Not observed

20.

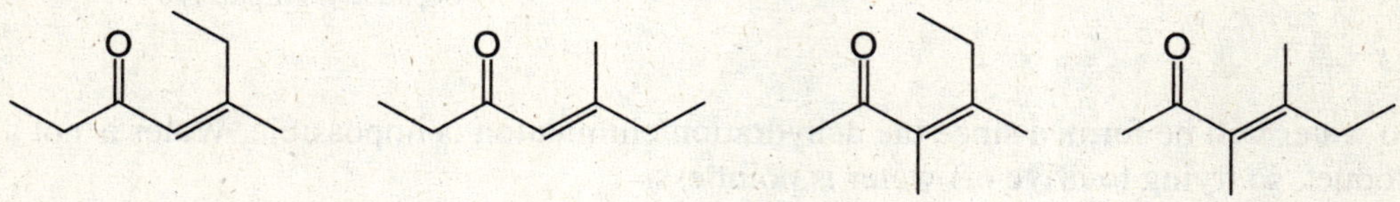

21. In base at low temp., the reaction stops at the aldol product. This reaction is overall uphill so the amount of product is very modest. With acid catalysis, even at low temp, the reaction proceeds all the way to the α,β-unsaturated ketone. This step from aldol to α,β-unsaturated ketone is downhill, making the reaction overall downhill, and giving a much better yield of product.

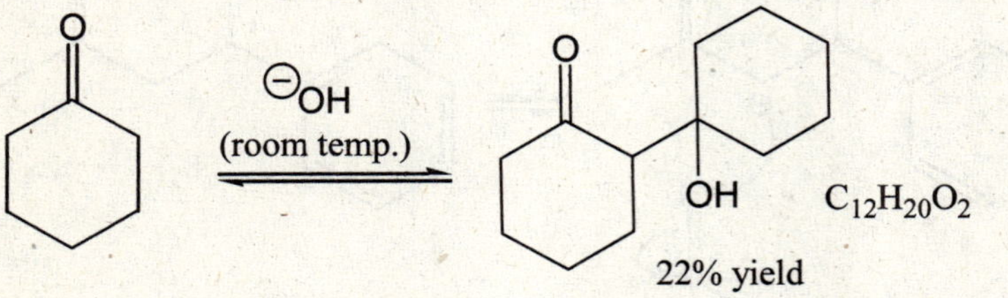

$C_{12}H_{20}O_2$

22% yield

At equilibrium there is mostly starting material

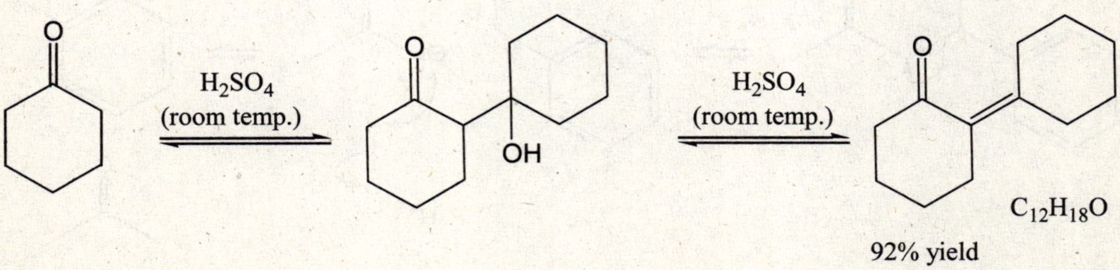

$C_{12}H_{18}O$

92% yield

22. a.

[aldol (not observed)] — [E] — [Z]

b. In this case, the elimination is very favorable because formation of the C=C allows the aromatic ring to conjugate with the carbonyl pi bond. This favorable interaction drives the dehydration and makes isolation of the aldol product nearly impossible.

23. a.

benzaldehyde + 2,4-dimethyl-3-pentanone →(hot concentrated base or acid) product — can not eliminate OH because no alpha H's

b. No water will be formed since the dehydration/elimination is impossible. Water is not a product, so trying to drive off water is pointless.

24. Mechanism of a reverse aldol.

25.

26.

27. The driving force for these rearrangements is the lowering of energy due to conjugation.

+ two other res. struct's

you can draw the 2nd order res. structure if you wish.

28.

For the synthesis below, the target shown in the question (an imine) is an error since amines preferentially add to the beta position of an α,β-unsaturated carbonyl compound.

ChemActivity 13: Claisen & Michael Reactions

Extend Your Understanding Questions (to do in or out of class)

16. Only the first ketone below can be made starting from ethyl acetate using the strategy in Model 4. The second structure cannot because a phenyl group cannot be added using a primary alkyl halide; the third structure cannot because it is not a methyl ketone; the fourth structure cannot because an isopropyl group would be added using a secondary alkyl halide (not primary); and the last structure cannot because the benzene ring could not have been added to the alpha carbon because the alpha carbon is part of the ring. (For this last one, there are many possible explanations for why it could not be made via the strategy in Model 4.)

17. See below.

18. See below.

19. Add methyl propionate slowly to a mixture of methyl benzoate and potassium methoxide.

20. The new carbon-carbon bond in the Michael product is marked below.

21. In a Michael reaction, the Michael donor acts as the nucleophile, after deprotonation.

Confirm Your Understanding Questions (to do at home)

22. This ester does not undergo a favorable (overall downhill) Claisen reaction because in the initial dicarbonyl product there are no H's alpha to both carbonyl groups that can be removed to make a true Claisen product. Only when there is such an H to be removed is the reaction downhill, since this allows the high energy alkoxide base to be transformed into a lower energy, highly resonance stabilized enolate base.

23.

a.

Continued on next page.

b.

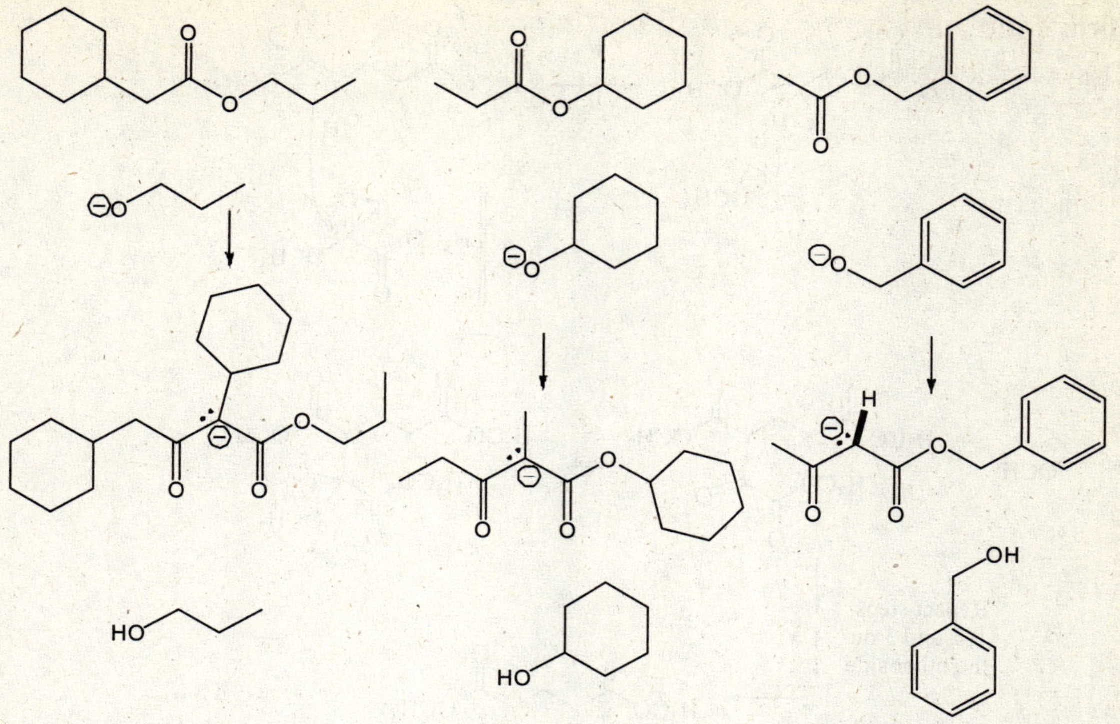

24. a)

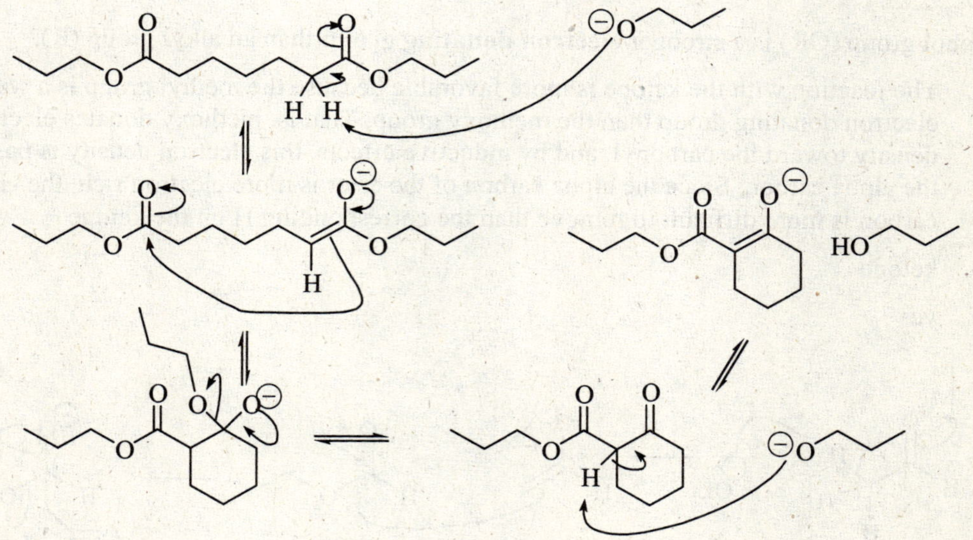

Continued on next page.

b) [reaction scheme]

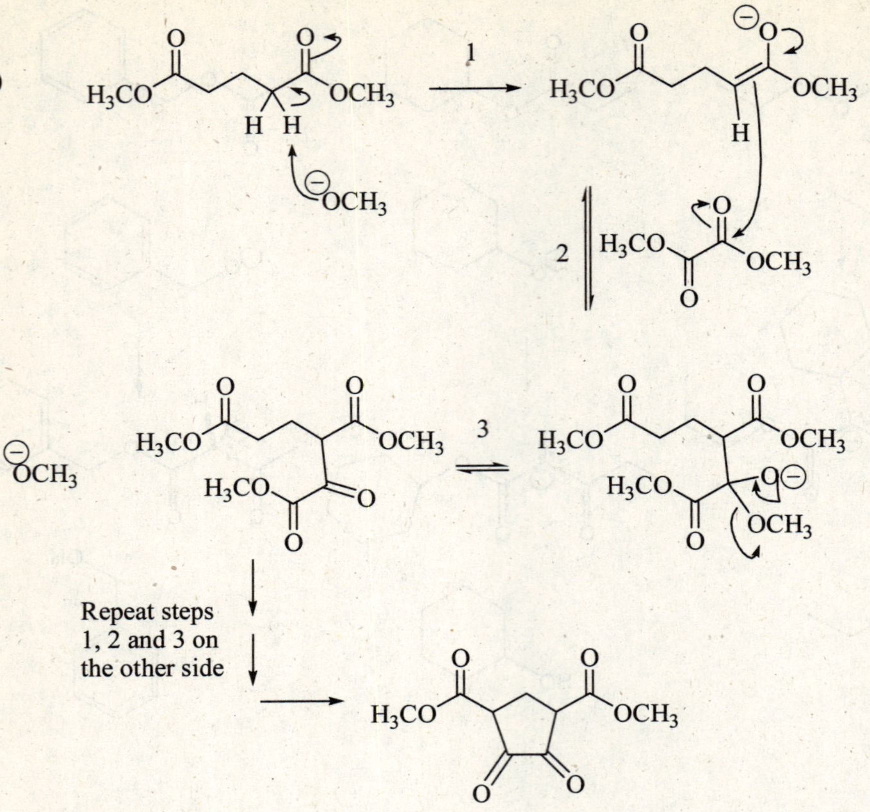

25. An alcohol group (OR) is a stronger **electron donating group** than an alkyl group (R).

    a. The reaction with the ketone is more favorable because the methyl group is a weaker electron donating group than the methoxy group. That is, methoxy donates electron density toward the carbonyl, and by inductive effects, this electron density is passed onto the alpha carbon. Since the alpha carbon of the ester is more electron rich, the H on this carbon is more difficult to remove than the corresponding H on the ketone.

    b. ketone

    c. yes

26. a. [mechanism scheme]

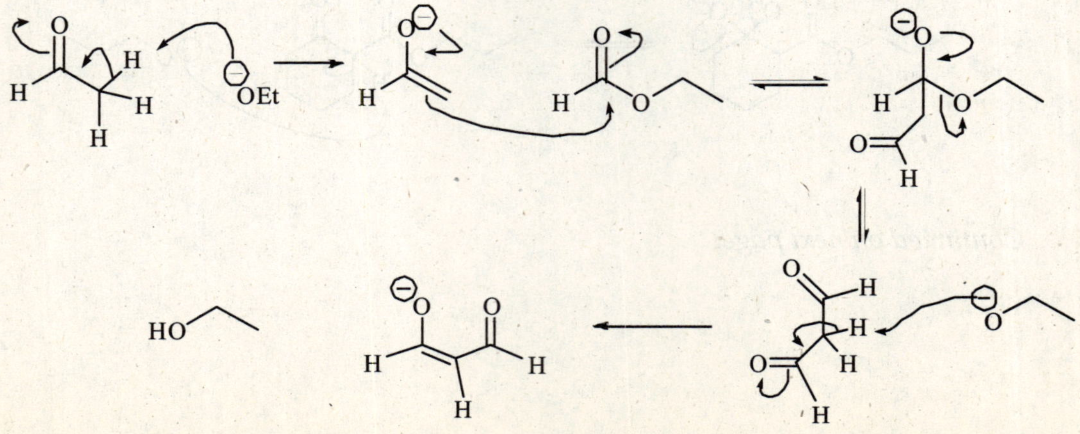

b. [structure: HO-CH(CH3)-CH2-CHO]

c. You could maximize formation of the Claisen product and minimize formation of the aldol product by slowly adding the nucleophile (acetaldehyde) to a mixture of the ethyl formate and ethoxide.

27.

[Reaction 1: methyl benzoate + acetaldehyde, ⁻OCH₃ → 1 aldol; 1 Claisen-type product (PhCO-CH⁻-CHO)]

[Reaction 2: methyl benzoate + methyl formate, ⁻OCH₃ → NO REACTIONS (no alpha H's!!)]

[Reaction 3: methyl formate + 2-butanone, ⁻OCH₃ → 2 aldol (since there are two kinds of alpha H's); Two different Claisen-type products]

[Reaction 4: methyl acetate + methyl propanoate, ⁻OCH₃ → Two different Claisen products]

[Boxed structures: two products with quaternary alpha carbon]

In the last reaction mixture the two products in the box will not form. The reason for this is that the final DOWNHILL deprotonation step drives the reaction. Neither product in the box has an acidic hydrogen that is alpha to both carbonyls. Without the possibility of this downhill deprotonation step, the reaction does not go to completion.

28. *See above*

29.

30.

31.

32. A Michael addition is like an aldol reaction in that the nucleophile is often an enolate. It is different from an aldol reaction in that the electrophile is (instead of an aldehyde) an alpha,beta-unsaturated aldehyde. Furthermore, an aldol product is a 1,2-addition product, while a Michael product is a 1,4-addtion product. That arises because, enolate nucleophile prefers to undergo 1,4-addition rather than 1,2-addtion, so when the electrophile is capable of both then the 1,4-addition product is likely. When no 1,4-addition product is possible, the enolate bonds to the carbonyl carbon leading to an aldol product.

33.
   a. The beta carbon is electrophilic.
   b. Yes

34.

35.

36. At normal concentrations, the intramolecular coupling with the other end of the dibromide is greatly favored over a coupling with a second dibromide since the former, by being attached already, is many thousands of times more likely to collide with the alpha carbanion. In general, intramolecular reactions involving transition states of five, six, or seven atoms (and sometimes larger numbers), are favored over intermolecular reactions.

37.

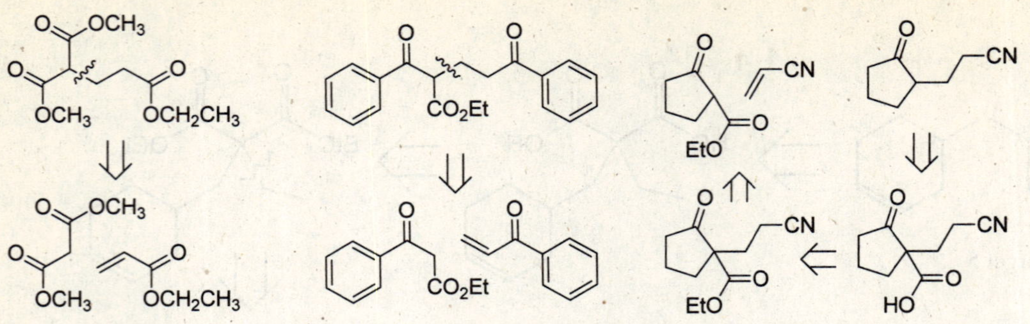

38. The first part of a Robinson annulation is a Michael reaction, the second part an aldol reaction.

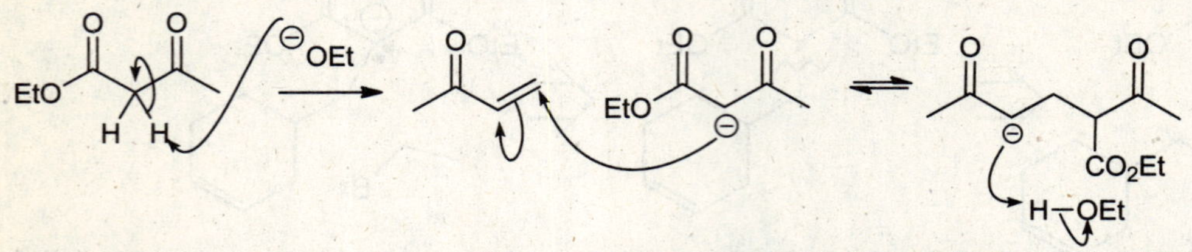

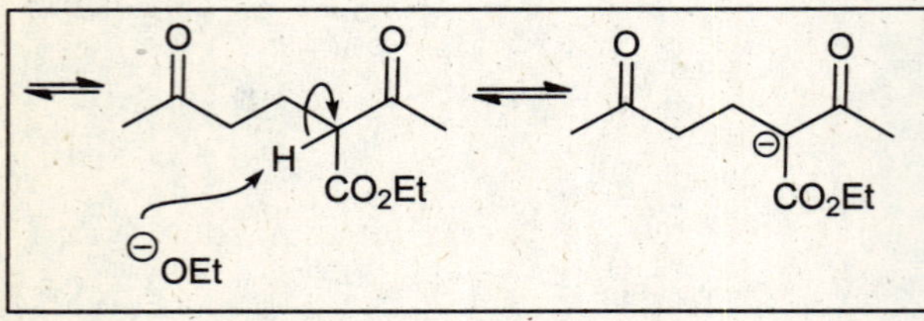

Though this H is the easiest to remove, this pathway can only lead to a four-member ring, which is not favorable. Though this enolate forms, it does not react further, but instead, reverts back to the previous compound.

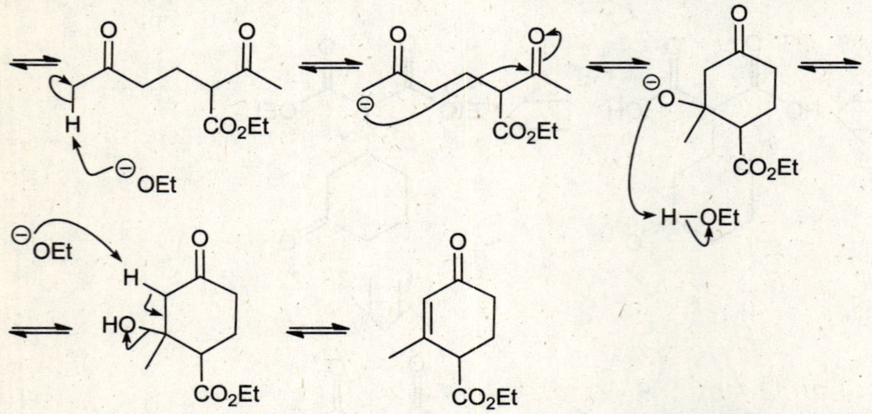

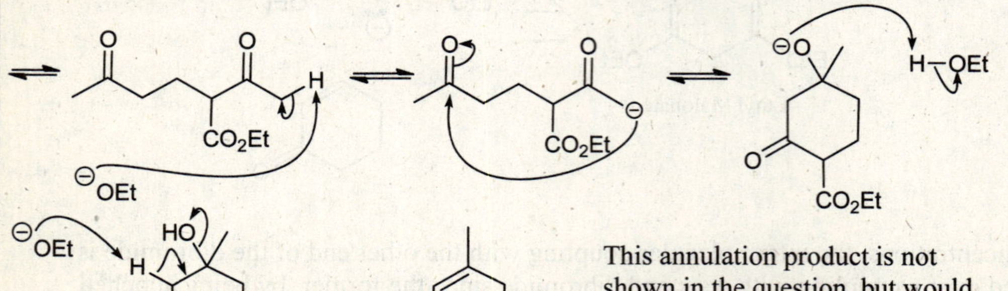

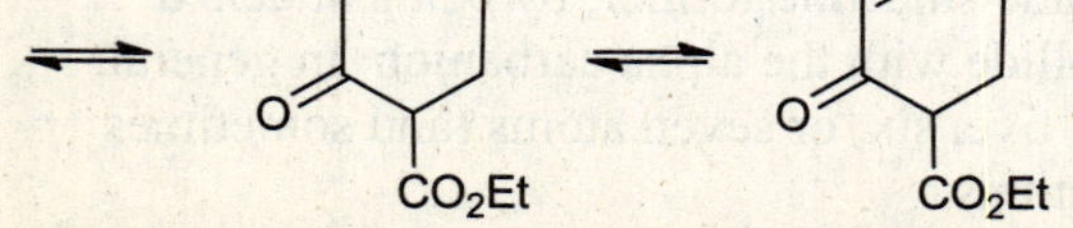

This annulation product is not shown in the question, but would also likely occur along with the product shown above.

ChemActivity 14: Amines

Extend Your Understanding Questions (to do in or out of class)

10. One way is to do a Gabriel synthesis using benzyl bromide as RBr. Another is to mix benzaldehyde and ammonia, then reduce the resulting imine using catalytic hydrogenation.

11. In the first box, the carboxylate ion will dominate at pH 7. For the second box, the ammonium ion will dominate at pH 7.

12. a. The cationic form of an amino acid (shown in Model 4) is expected to dominate below pH 5.
    b.
    c. (Check your work.) In each box on the graph below, draw the form of an amino acid that is expected to dominate over the pH range covered by the box.

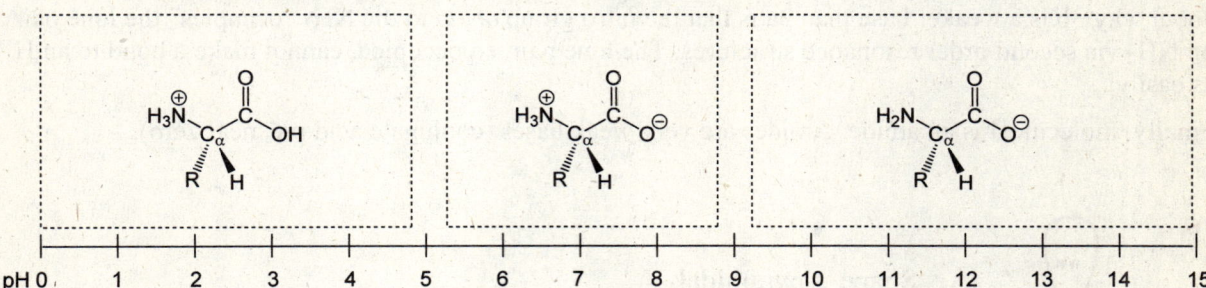

Confirm Your Understanding Questions (to do at home)

13. Delocalization of the lone pair on N makes this lone pair less likely to form a bond to H than the lone pair on a non-aromatic amine such as cyclohexylamine. The likelihood of a lone pair on a base to form a bond to H⁺ is the definition of base strength, so aromatic amines are much weaker bases than aliphatic (non-aromatic) amines.

2nd order resonance structures of aniline

14. Molecules marked 1 and 2 are stronger bases than aniline. Molecules marked 3 and 4 are weaker bases than aniline.

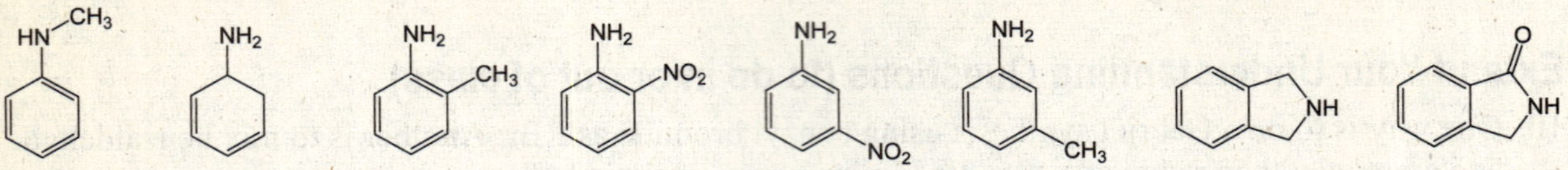

2a, 1b, 2b, 3b, 3a, 2c, 1a (strongest base), 4 (weakest base)

15. It is difficult to predict the exact order of base strength, but the numbers above define four groups of molecules of decreasing base strength.

1a and 1b are the strongest bases because they are non-aromatic amines (conjugate acid pK_a around 9), and 1a is probably stronger since it is secondary (and alkyl groups are generally electron donating).

2a-c are the next strongest bases. Though they are anilines (conjugate acid pK_a near 5), they each have an electron donating alkyl group. This makes the system more electron rich, and therefore makes the molecule more basic (than aniline). 2a is hard to compare with 2b and 2c since it is hard to compare the effects of an alkyl group directly attached to N, versus on the ring. Between 2b and 2c it is likely that 2b is the stronger base since the effects of a substituent on an aromatic ring are focused on the *ortho* and *para* positions. This means that the methyl group *meta* to the NH2 on 2c would have less of an electron donating effect than the methyl group *ortho* to the NH_2 on 2b.

Using the same argument, 3a is expected to be a stronger base than 3b. This is because the nitro group withdraws electron density, making the system less electron rich, and a weaker base. One way to think about why 3b is a weaker base than 3a is that the nitro group *ortho* to the NH_2 "occupies" the lone pair on NH_2 via second order resonance structures. The lone pair, so occupied, cannot make a bond to an H^+ as easily.

Finally, molecule 4 is an amide. Amides are very weak bases (conjugate acid pK_a near zero).

16.

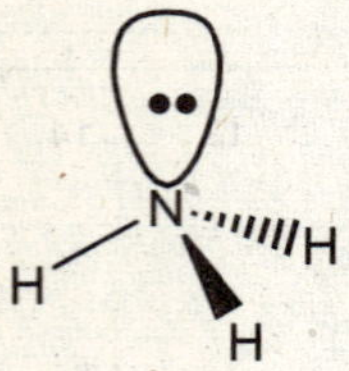

Shape = pyramidal
Hybridization state = sp^3

17. An amine is constantly "flipping" back and forth like an umbrella inverting and un-inverting in a strong wind, as shown below. Addition of acid can "lock it" in either position. Under normal conditions, this will give rise to a racemic mixture of the R and S salts.

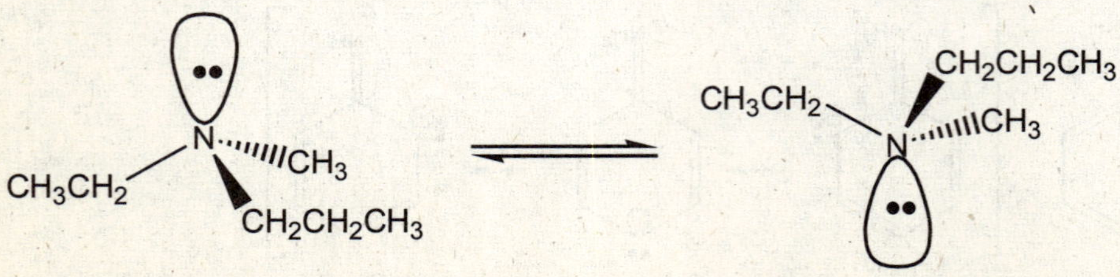

Continued on next page.

a. **R** enantiomer of the conjugate acid of *N*-ethyl-*N*-methylpropylamine.

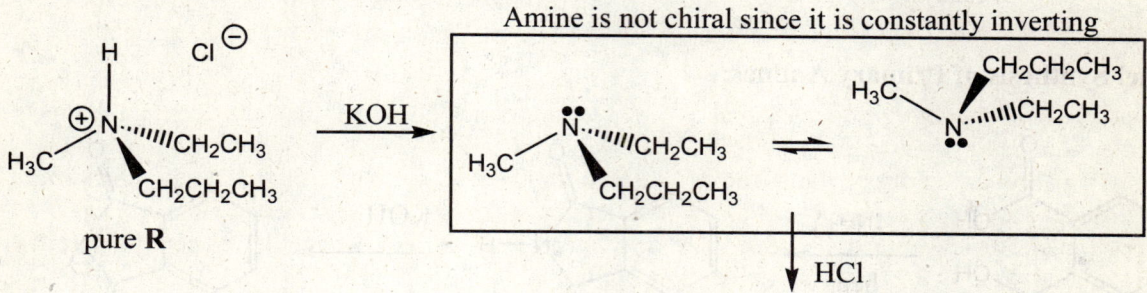

(**R**)-*N*-ethyl-*N*-methyl propylammonium chloride

(**S**)-*N*-ethyl-*N*-methyl propylammonium chloride

b. It will result in a racemic mixture since the free amine will racemize.

18.

19. *see above*

20. Pyridine (the structure on the left) is more basic. The lone pair on the N of pyridine is localized on N and NOT involved in the aromatic pi system. The lone pair on the N of pyrrole *is* involved in the aromatic pi system. This means it is delocalized. Another way of looking at this is…if the N of pyridine picks up an H⁺ it does not have an impact on the aromatic pi system, but if the N of pyrrole picks up an H⁺, the result is complete loss of aromaticity. Therefore pyridine is more likely to act as a base (i.e. pyridine is a stronger base).

21. Three different ways of making benzyl amine (see Q10) using ammonia (NH_3).

PhCOCl →[NH_3] PhCONH$_2$ →[$LiAlH_4$] PhCH$_2$NH$_2$

PhCHO →[NH_3, acid catalyst] PhCH=NH (imine) →[H_2/Pd] PhCH$_2$NH$_2$

Gabriel Synthesis of Primary Amines:

phthalic acid →[:NH_3, heat (acid catalyst), $-2\,H_2O$] phthalimide (N–H) →[KOH] phthalimide anion (:N:⁻) →[any 1° alkyl halide, Br–CH$_2$Ph] N-benzyl phthalimide + Br⁻ →[H_2SO_4, H_2O] phthalic acid + PhCH$_2$NH$_3^+$ →[neutralize w/ base] H_2N–CH$_2$Ph

22. In what pH range (if any) is an amino acid likely to...

   a. at no pH!

   b. in the middle range shown below, from about 5 to 9.

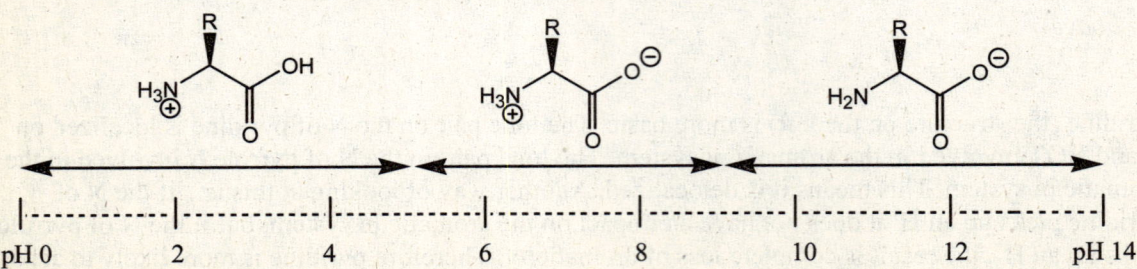

Continued on next page.

Note that the answers above assume that the two functional groups on the backbone of an amino acid do not communicate with one another. In fact, they do. For example, at acidic pH the $^+NH_3$ group acts as an inductive withdrawing group making the carboxylic acid more acidic. This means that the zwitterion range starts much lower than pH 5, more like pH 3.

23. Mechanism of a Gabriel synthesis (See Model 3.).

Continued on next page.

152 Volume 2, ChemActivity 14: Amines

mechnism is the reverse of the twelve steps shown above

24.

Second order resonance structure

a. See above.

b. The second order resonance structure of the starting amide (shown above) tells us that there is a partial + charge on N in the starting material and that the lone pair on N is delocalized into the carbonyl. The amine has no comparable second order resonance structure. This means the reaction has made the N much more nucleophilic (and more basic). Recall that nucleophilicity depends mostly on two factors: base strength and size.

c. See above.

25. Biologists often refer to the carbon marked with an alpha, below, as the alpha carbon because it is alpha to the carbonyl.

26. Nineteen of the twenty commonly occurring amino acids are chiral. Of these, all but one has an absolute configuration of **S**. Only cysteine (R = CH_2SH) has an absolute configuration of **R**.

27. Cysteine has a different absolute configuration even though, like the others, it has an H coming out of the page and the R group going into the page (with the amino terminus on the left) because of the priority system used. In short, sulfur beats an oxygen in this system, which gives this side-chain priority over the carboxylic acid group.

Nomenclature Worksheet 3: Naming Benzene Derivatives

9.

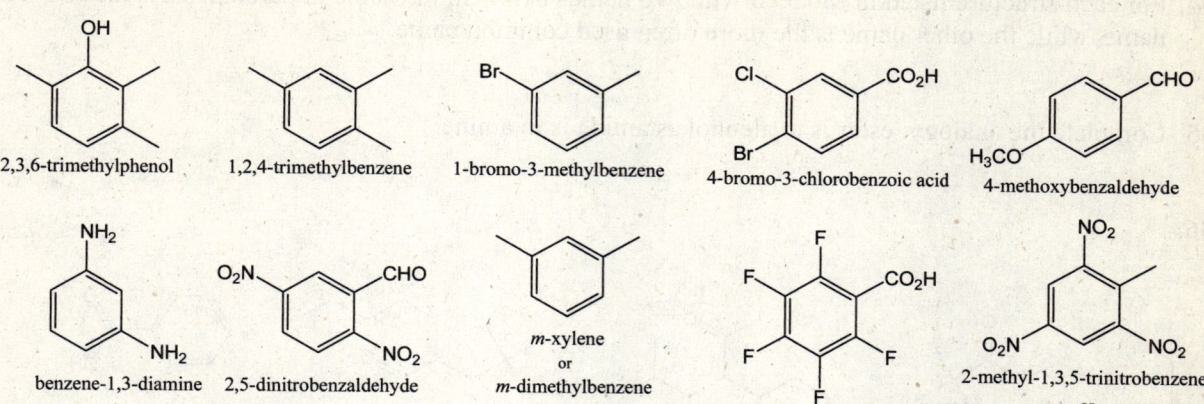

10.

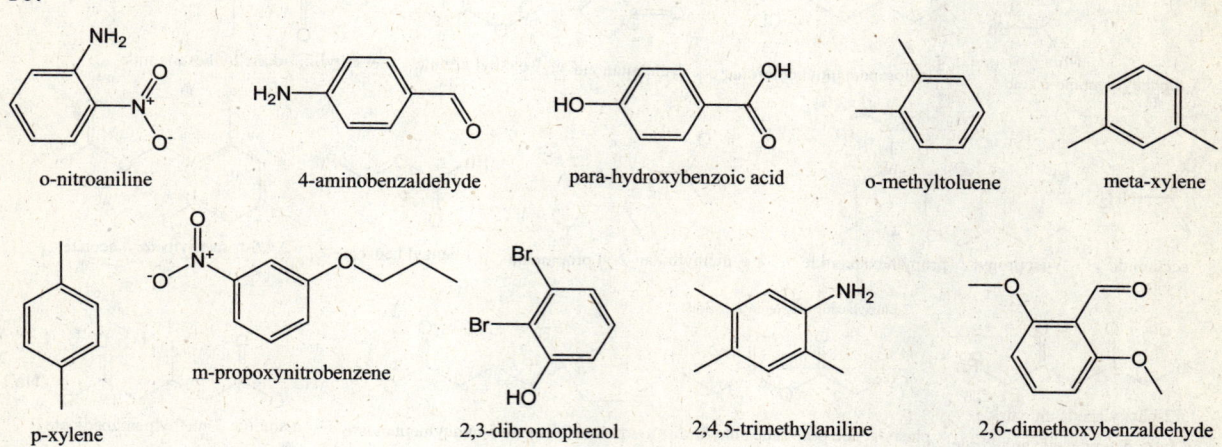

Nomenclature Worksheet 4: Naming Carbonyl Compounds

34. For each structure listed in Model 3 with two names below it, the name in parenthesis is the IUPAC name, while the other name is the more often used common name.

35. Complete the analogy: ester is to alcohol as amide is to amine.

36.

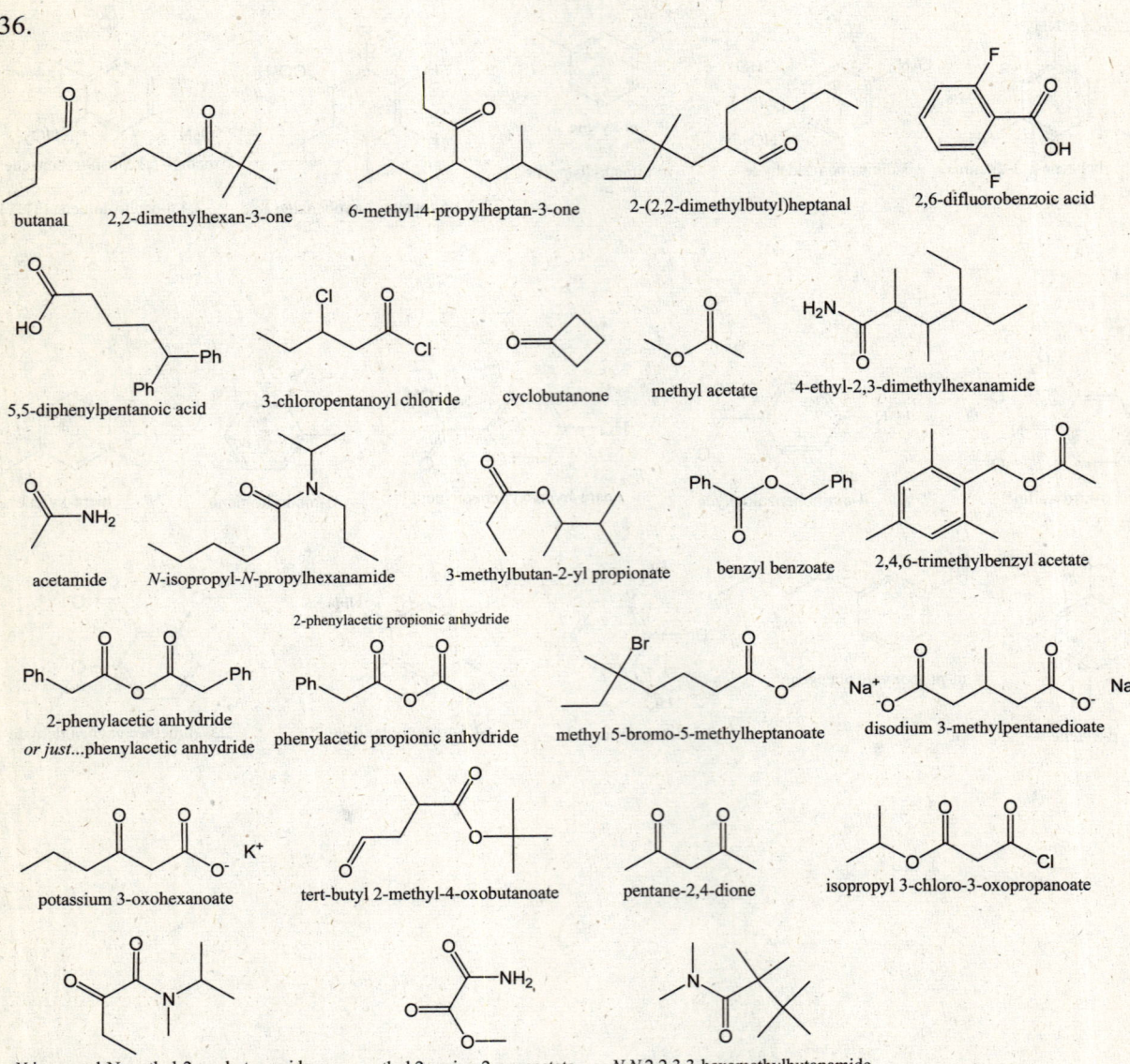